THINKING TOWARD SOLUTIONS:
Problem-Based Learning Activities for General Biology

Student's Manual

DEBORAH E. ALLEN &
BARBARA J. DUCH

Department of Biological
Sciences and Mathematics
and Science Education
Resource Center
University of Delaware

Saunders College Publishing

Harcourt Brace College Publishers

Fort Worth Philadelphia San Diego New York Orlando Austin
San Antonio Toronto Montreal London Sydney Tokyo

CONTENTS

Part Three:

FORMS TO HELP YOUR GROUP FUNCTION WELL *149*

Part Four:

SOME HELP WITH SPECIAL ASSIGNMENTS *157*

Part Five:

BIBLIOGRAPHY *161*

Acknowledgments

We wish to acknowledge the financial support for the problem-based learning effort at the University of Delaware by the National Science Foundation's Division of Undergraduate Education and the U.S. Department of Education's Fund for the Improvement of Post-Secondary Education (FIPSE). We thank the many colleagues with whom we have collaborated over the past four years in developing ways to incorporate problem-based learning strategies into undergraduate courses, and the University administration for their continuing support of this project.

We would also like to acknowledge the capable assistance of Jennifer Johnson with locating Internet and other resources for researching the problems. And to the students and undergraduate peer tutors who have participated in our problem-based learning courses, we offer our warmest thanks for freely giving the feedback on the problems and the structure of the courses that continues to help us make them better.

INTRODUCTION:

WHAT IS PROBLEM-BASED LEARNING?

The basic principle supporting the concept of problem based learning (PBL) is older than formal education itself: namely, that learning is initiated by a posed problem, query, or puzzle that the learner wants to solve (Boud and Feletti, 1991). In the problem-based approach, complex, real-world problems are used to motivate students to identify and research concepts and principles they need to know in order to progress through the problems. In a problem-based class, students work in small learning teams, bringing together their skills at acquiring, communicating, and integrating information in a process that resembles that of scientific inquiry.

Problem-based instruction addresses many of the recommended approaches that will help you become successful in college and in your working career after college. Leaders in business and industry believe that those skills are the following:

- To think critically and be able to analyze and solve complex, real-world problems
- To find, evaluate, and use appropriate learning resources
- To work cooperatively in teams and small groups
- To demonstrate versatile and effective communication skills, both verbal and written
- To use content knowledge and intellectual skills acquired as a college student to become a continual learner

A problem-based course may be structured in a variety of ways based on choices that your instructor makes. However the general process of problem-based instruction (Boud and Felletti, 1991) is the following:

- Students are presented with a problem (case, research paper, and videotape, for example). Students in their working groups organize their ideas and previous knowledge related to the problem, and attempt to define the broad nature of the problem.
- Throughout discussion, students pose questions, called "learning issues," on aspects of the problem that they do not understand. The group records these

learning issues. Students are continually encouraged to define what they know—and more important, what they don't know.

- Students rank, in order of importance, the learning issues generated in the session. They decide which questions will be followed up by the whole group, and which issues can be assigned to individuals, who later teach the rest of the group. Students and instructor also discuss what resources will be needed to research the learning issues, and where they can be found.
- When students reconvene, they explore the previous learning issues, integrating their new knowledge into the context of the problem. Students are also encouraged to summarize their knowledge and connect new concepts to old ones. They continue to define new learning issues as they progress through the problem. Students soon see that learning is an ongoing process, and that there will always be (even for the teacher) learning issues to be explored.

This PBL cycle is therefore designed to help you develop the ability to identify the information that is needed for a particular application—where and how to seek it, how to organize it in a meaningful conceptual framework, and finally, how to communicate this information to others.

PART I

WORKING IN GROUPS

An important element in your success in a problem-based course is your assigned group and how it functions. Use of cooperative working groups in a science class, even in large enrollment ones, helps to foster the development of learning communities and lessens the sense of isolation that students might otherwise feel. Research has shown that student achievement is enhanced when students work together in a cooperative learning environment (Bodner, 1992; Johnson et al., 1991), and these students earn higher grades than students who try to learn the same material individually (Johnson et al., 1991). Cooperative learning also helps to increase the motivation to learn and the interest to solve more complex problems (Johnson et al. 1991). In addition, social and team skills learned in student groups are important for success in the working world today.

Some Common Questions about Working in Groups

If this is your first time to work in a group, or even if you have had many previous experiences, you will probably have questions or reservations about the process. The following questions and answers might address some of your concerns.

Question 1: I don't like to do all the work and let others take the credit. How can I prevent this?

It is important that each student be held individually accountable for his or her own performance. When each student group takes this seriously, "free riders" are discouraged and contributors to the group effort are rewarded. The first provision for emphasizing individual accountability is the assignment of roles of responsibility. It works well to have a separate role for each student in your group. These roles rotate among group members every week or after every problem or assignment so that each group member has a chance to practice the skills that performing each role entails. Some suggested roles and their responsibilities are included in Table 1.

Table 1. Roles and responsibilities of group members

Role	Responsibility	For Example
Discussion Leader	Keeps group on track; maintains full participation	"Let's focus on the problem." "Should we move onto the next question?"
Recorder/record keeper	Records assignments, strategies, unresolved issues, data; convenes group outside of class; keeps group record sheets	"Did we get all the learning issues down?" "Is this the diagram we want?"
Reporter	Makes sure all agree on plan/action/report; writes up assignment	"Are we in agreement now?" Everyone check this draft before tomorrow."
Accuracy Coach	Probes for group understanding; locates resources and brings them to class	"Why do you think that?" "What does this text say?" "Where did you find that information?"
Skeptic	Challenges group consensus; checks for alternative ideas	"I'm not sure we're on the right track." "Are you sure about that? John, do you agree with Sue?"
Timekeeper	Checks for timing of discussion of the problem; alerts discussion leader when group needs to switch activities	"Lisa, we've only got 10 minutes, we'd better assign learning issues." "We should move onto the second page now."
Reflector/ Summarizer	Summarizes progress of the group; checks for involvement and understanding of all group members	"Here's where I think we are. What do you think, Sally?" "Jack, you haven't said what you're thinking. Do you agree with Joe?"

If there are fewer students in your group than roles of responsibility, you may want to combine some of the functions, such as discussion leader and timekeeper, or accuracy coach and reflector.

Another important tool that will help group members ensure that all students contribute to the group effort is establishing and enforcing procedural guidelines or group ground rules. You should seriously consider the behaviors that you will and will not tolerate from one another, and decide the consequences for violators of these ground rules. It is important to make these decisions in the first week or two of class before negative behavior begins. After your group comes to consensus on the ground rules and consequences, your instructor will request that you sign two copies—one for yourselves and one for your instructor.

You might begin your discussion about the types of rules you want by talking about past experiences with groups, listing both the positive and negative aspects. Then think about ground rules that will reinforce the positive behavior and limit the negative. Some examples of commonly used ground rules are the following:

- Come to class on time every day
- Do all assignments and be prepared to discuss them
- Notify group members of absences
- Share information
- Respect the views and ideas of others
- Reporter should share a draft of the write-up two days before it is due

As in the world outside your classroom, rules need to be backed up with consequences, or they may get ignored. Some examples of appropriate consequences for violators of the ground rules are the following:

- Group member will have a "time-out" period from the group and will be responsible for completing all work on his/her own
- Group member will not receive a grade for an assignment for which s/he did not contribute
- Group member will be responsible for a greater share of the next assignment
- After two ground-rule violations, a member can be expelled from the group permanently

Be sure to have your instructor approve your list of ground rules as well as consequences, so that if needed, s/he will help you reinforce those guidelines.

Question 2: I've been in groups before, and I don't like being slowed down by other group members. How can I change this? Or I'm not really good in this subject, and I'm afraid I'll hold back my group.

One aspect of cooperative learning that helps maximize every student's learning is explaining material to other group members. Every teacher will tell you that it isn't until they have actually taught a topic that they have developed a deep level of understanding it. When students teach other students they reinforce their own understanding of material, and are forced to face elements of a concept that might not be clearly understood. Many times students can understand another student's explanation because that other student has also just learned the material and can clearly understand the conceptual pitfalls. So everyone is a winner in a cooperative group—the one who instructs and the one who listens; both students learn.

Question 3: What can I do to get group members to do their assignments?

If your group has written thoughtful ground rules and consequences, then the next step that will help you reinforce positive group behaviors and maximize individual accountability is giving helpful feedback to individuals and the group as a whole. Group feedback sessions should be scheduled two to three times a semester or whenever you perceive that the group is not functioning at its best. The feedback session could begin with each individual stating what the group did well since the last feedback session, and what each student thinks the group needs to change or improve in order to function better. You may find the form "Prompts for Discussion of Group Function" in Part III convenient to use to help your discussion begin. After the recorder lists all suggestions, the group can go on to discuss the feedback, then come to consensus on one or more goals for the group over the period of time until the next feedback session. For example, if some members of the group are not coming prepared to class, and this is leading to a breakdown in group discussion, the goals for the next two weeks could be one or more of the following:

• Every member will arrive on time for class, prepared to discuss his/her assignment.
• The discussion leader, accuracy coach, and skeptic (or reflector) will challenge each individual to contribute during discussions and will discourage any splintered discussion.
• Each member will submit his/her assignment in writing as a part of the group recordkeeping.

After you have discussed the current functioning of your group and set goals for future interactions, you are then prepared to rate each individual's contribution to the group and write feedback for each group

Table 2. Sample feedback

Student Behavior	Sample Feedback
Student A is quiet and doesn't contribute to discussion	"I think it would help the group if you'd contribute to the discussion more often. The information and opinions that you do contribute are very knowledgeable and helpful to our learning."
Student B dominates the discussion and is frequently incorrect.	"I've noticed that you contribute to the discussion, but often don't allow others to speak. I think this hurts the group when your information may not be correct or complete, and others in the group don't have a chance to add their knowledge. I feel that our group will be better if everyone has an opportunity to share what they know."
Student C thinks s/he knows the material and tries to get the group to move quickly through the problem, getting impatient if anyone asks questions.	You're a good resource of knowledge for the group, but when you encourage us to move too quickly through the problem, we don't learn all the details that we need. It would be helpful to the group if you asked more challenging questions, and helped answer others' questions."

member on a feedback form (available in Part III). The feedback form is filled out in a confidential manner and given to your instructor, who will summarize the results and return it to each student. The results of the ratings will be factored into your grade (as detailed in your course syllabus).

Giving constructive feedback to members of your group is an important skill that you need to develop. It is also a powerful tool in managing the effectiveness of your group. Table 2 gives some examples of student behavior and some suggested feedback statements.

It is worth pointing out that the examples of feedback provided in Table 2 fall within the guidelines for giving good feedback that many experts recommend (Gibbs, 1995). Some of the "rules" for constructive feedback that are most relevant to the process described here are as follows. Good feedback:

- Describes behavior rather than criticizing or demeaning it
- Gives specific (rather than general) information about the behavior
- Is presented as an individual's perceptions or feelings rather than as absolutes
- Focuses on behavior and actions rather than on someone's personality
- Focuses on behavior that can be controlled or changed

Question 4: How can a group be fair about dividing responsibility in a big project?
It is important to divide responsibilities among your group members when you have been assigned a large project. At each stage of the assignment, clearly document in writing each person's responsibility and the due date in your record-keeping sheet (see "Checklist for Problem-Solving" in Part III). In fact, you will find that record keeping is a good way to eliminate conflicts, tensions, and disagreements about the division of labor associated with all problems, cases, or assignments.

Question 5: I'm trying to get into grad school, so grades are important to me. What if I'm in a group with folks that don't care about getting good grades?
Group ground rules, roles of responsibilities, documenting the activities of the group and individuals within the group, and peer pressure within the group help raise expectations of all group members. Each student is responsible for monitoring the functioning within his/her group and the academic level of the discussions, assignments, and research reported. If you are in a group with a person who resists working as hard as the rest of the members, you need to use all of the group-monitoring methods discussed here, giving that individual truthful and direct feedback on his/her performance and the group's expectations. As a last resort, you may wish to seek advice from your instructor, who will help you find ways to deal with conflict and will be prepared to intervene directly, if necessary.

PART II

USING THE PROBLEMS

The problems in this book are different from the typical problems and end-of-chapter questions that you find in biology textbooks. These problems are contextually rich, and related to real-world situations that are connected to the biological principles you are learning. Studies have shown that when learning is focused around realistic and open-ended situations, students are more likely to retain what they learn and apply that knowledge appropriately (Albanese and Mitchell, 1993; Boud and Feletti, 1991).

To work through this type of problem, you need to develop and use a variety of skills that may be new to you. This section presents a general outline of the process that you and your group could use to solve an assigned problem.

Problem-Solving Process

When you and your group begin a problem, it's a good idea to first identify the broad nature of the problem, listing the major concepts or issues that need to be addressed. The discussion leader could lead a brainstorming session, in which each individual discusses what s/he already knows about the topics. As you continue discussions, some questions and issues will arise that you don't know or don't understand. A useful role for the recorder at this point is to keep a running list of these items that we generally call "learning issues." In many cases, you may have a learning issue that another member of your group can explain so that the group can move forward. Toward the end of the first session, it's important to prioritize the list of questions and learning issues, listing the "big ideas" or larger conceptual issues first and definitions and vocabulary questions last. This will help your group to decide the best order of presentation of researched issues when you come back to the problem, and to connect and synthesize what you've learned. The group should then decide which learning issues should be researched by everyone in the group (usually the central idea[s] in the problem) and which can be assigned to individuals. To help with ensuring even distribution of the workload, the recorder or record keeper should document each person's assignment. Before leaving class, it is also a good idea to discuss what learning resources will be useful for researching an issue and where these resources can be located (e.g., library, Internet, course textbook). The Checklist for Problem-Solving found in Part III is designed to help you monitor your progress on the problem and document the assignment of learning issues as you go through this process.

In your next class session, each group member needs to come prepared to teach the other group members about the learning issue(s) s/he researched. This activity is crucial in helping you to hone your communication and questioning skills. This is also an opportunity for revisiting a discussion or brainstorming session based on the new information learned, and allows the group to reach a deeper understanding of the content issues. You are then ready to move on to the second stage of the problem, and the learning cycle will then repeat itself for this and any subsequent stages.

Stages of Problems

Notice that many of the problems have more than one stage. This is a common practice in problem-based learning so that students can build on their growing understanding of complex issues without having to tackle them all at once. In many instances, you will want to bring early stages of the problem to the best possible resolution before moving on to later stages. Often, working on subsequent stages will refine your understanding of the earlier issues. Your instructor may bring the whole class together periodically to clarify ideas or make assignments to insure that you have reached a level of resolution that allows you to move to the next stage .

Finding Good Resource Materials

One of the common problems encountered by people researching the scientific literature for the first time is knowing how to narrow and refine the search for the most relevant and useful materials and resources. No one has the time—least of all a student who is taking many other courses—to spend endlessly in the library, only to "come up empty-handed." To help you learn the art of conducting a good search, two sections are included at the end of each problem, "Topics Introduced by the Problem" and "Additional Resources for Researching Problem Content."

The first of these sections is a list of topics that you will likely find in your biology textbook. Because these texts are well-written and well-suited to the beginning biology student, they are a good place to start your search. There is a

Table 3. Topics versus learning issues

Topic	Learning Issue
Cigarette smoking	Is nicotine addictive—who should we believe?
Genetic testing	What are the pros and cons of performing genetic testing on an unborn child (fetus)?
AIDS	How is HIV transmitted from one person to another?

separate list of more advanced topics not likely to be found in your textbook, but that may be necessary for researching your learning issues. To help you get started finding information on these topics and others that you may want to research, the second section provides some starting points in the library and the Internet. You no doubt will want to go beyond these initial suggestions as you delve deeper into the issues you have identified. It is a good idea to write down the sources you find so that you can share them with your group and other members of your class.

It is important to emphasize that a topic is not a learning issue. That is, by giving you this list of topics at the end of each problem, the process of identifying learning issues for yourself has not been short-circuited. You might wonder, "what's the difference between them?" A topic is a word or phrase indicating a general or specific content area in biology. In contrast, a learning issue is usually posed in the form of a question that gets at the heart of what you will need to know to make progress through the problem. Table 3 presents some examples of topics and learning issues that could be related to them.

The Problems

Stage 1:

DON TRIES TO CULTURE FISH CELLS

Deborah E. Allen

Don Tilapia is a recent graduate in biology from Ocean State University. He has just started a new job as a research associate at the university's aquaculture facility.

The manager of the facility is interested in setting up some fish cell lines for use as "biomarkers" for sub-lethal exposure to heavy metal toxins, as well as for examining the ways that cells metabolize and detoxify these substances. He wants to do this for several species: *Ictalurus nebulosus* (the brown bullhead catfish), *Scophthalamus maximus* (the turbot), and *Squatina dumerili* (the angel shark). However, just two days after Don started working, his boss (the manager) went on a two-week vaca-tion. Don decides to try a couple of his own ideas on getting cells of these species established in culture, hoping to pleasantly surprise his boss when he returns.

Don starts by harvesting some muscle cells from each species (he chooses these cells because he reasons that muscles are metabolically active cells). He then places the catfish cells in a chamber containing water from a local freshwater pond, which he first passes through a filtration system, and the turbot and angel shark cells in separate flasks containing filtered synthetic sea water.

The results are disastrous. All three cultures fail to become estab-lished, and the view he gets of his cultures by looking under a micro-scope is not an encouraging sight.

Questions to ponder:

- Where did Don go wrong in each of the three instances?
- What might Don have seen when he looked at his cultures under the microscope?
- What power objective lens did Don most likely use to try to visualize his cell cultures?

Stage 2:
DON TRIES TO CULTURE FISH CELLS

Not to be deterred, Don recalls that the lab manual from his college physiology course has the recipe for a fish Ringers solution. The ingredients

salt	concentration of salt in g/L
NaCl	6.42
KCl	0.15
$CaCl_2$	0.22
$MgSO_4$	0.12
$NaHCO_3$	0.084
NaH_2PO_4	0.06

are as follows:

"I'm set," thinks Don as he mixes up a liter of the solution, then tests its osmolality by running a small sample through a freezing-point depression osmometer. He comes up with a reading of 320 mOsm/kg H_2O.

Satisfied, he harvests cells from each species once again, and puts them each in separate flasks containing his new Ringers solution.

The next day, he is relieved to find that his catfish and turbot cells are still alive, but chagrined to find that his angel shark cells have not done too well, to put it mildly.

Questions to ponder:

- Why did Don's fish Ringers solution disturb the cells of one marine fish, yet allow for the (at least temporary) survival of those from another marine species?
- From Don's experiments with cell culture, what conclusions can you draw about the solute concentration of the immediate environment that surrounds each of these cells in vivo, versus the composition of the natural environment that surrounds the whole fish?
- Is 320 mOsm/kg H_2O a realistic value for the osmolality of Don's solution, assuming he made it up correctly? That is, what is the relationship between the salt concentra-

Stage 3:

DON TRIES TO CULTURE FISH CELLS

· ·

tion in g/L and the osmolality?

Don adjusts his angel shark tissue culture medium by increasing the concentrations of some of the ions, omitting others, and adding an important missing ingredient. After mixing up the solution, he takes three separate samples and measures the osmolality on his freezing point depression osmometer; he is satisfied with the resulting readings of 1020, 1025, and 1022 mOsm/kg H_2O.

Thinking he has the problem licked, he puts cells from each species into aerated solutions that provide the proper osmotic environment. Two days later, he is dismayed to discover that, once again, his cell cultures are not a pretty sight.

Fortunately, just before his boss is due back from vacation, Don runs across a few Web sites and books that help him figure out his problem.

"Culture fish cells? No problem..," or so he says to his boss as he greets him on his first day back to work.

Questions to ponder:

- What other factors did Don realize he needed to consider (other than ionic composition of the medium) to keep cells alive in vitro? Are these different from survival conditions of cells when they are part of an organism? Are they different from conditions needed by a paramecium (an organism that consists of a single cell)?
- What general recommendations about cell type might Don offer his boss (as being the best to culture for use in his toxicity studies)?
- What are the pros and cons of using cells to do the types of studies that Don's boss has in mind?

Topics Introduced by the Problem:
DON TRIES TO CULTURE FISH CELLS

Basic Background Topics

(Your textbook can be consulted for information on these.)

• Osmosis
• Cell membrane permeability and selective permeability
• The importance of maintaining the cellular osmotic environment in a relative state of constancy
• The requirements of life at the cellular level

More Advanced Topics

(You may want to consult additional resources for information on these.)

• Cell culture techniques
• Osmoregulatory strategies used by fish
• Composition of body fluids in fish/vertebrates

Additional Resources for Researching Problem Content:
DON TRIES TO CULTURE FISH CELLS
••

Books:
Darling, D. C., and S. J. Morgan. 1995. *Animal Cell—Culture and Media.* Boston: John Wiley and Sons.

Freshney, R. I. 1993. *Culture of Animal Cells—A Manual of Basic Technique.* Boston: John Wiley and Sons.

Martin, B., 1997. *Tissue Culture Techniques: An Introduction.* Boston: Birkhauser.

Electronic resources:
American Type Culture Collection Home Page. <http://www.atcc.org/>

Fletcher, M. "The Use of Fish Cell Culture Preparations in Environmental Toxicology." <http://www.science.mcmaster.ca/Biology/4S03/MF2.html>

The Salk Institute for Biological Studies. *Bart's Cookbook and Lab Protocols.* <http://flosun.salk.edu/users/Sefton_web/Hyper_protocols/TableOfContents.html>

Stage 1:

THE CURSE OF THE MUMMY
(If It's Dead, It Sphinx)

Jane Noble-Harvey

Professor Noseitall from the University of Delaware and his associate, Dr. Schmart, have received a grant to investigate the recently discovered tomb of Egyptian Pharaoh Hotsitotsi III, who ruled Egypt from 1542 B.C. until 1520. B.C..

The outer rooms of Hotsitotsi's tomb have been opened by the Egyptian government, but the burial chamber itself remains sealed, and has been undisturbed for more than 3,500 years.

Noseitall and Schmart make extensive preparations for entering the burial chamber and analyzing its contents. They have very limited information about such unopened chambers. They know the temperature will be cool, around 55 degrees Fahrenheit, and the atmosphere will be exceptionally dry. These conditions are responsible, in part, for the preservation of the materials in such chambers.

Questions to ponder:

- What infectious agents do you think the scientists might encounter in the tomb that would still be capable of infecting humans?
- How do infectious agents get into humans?
- How do you think the scientists might protect themselves from infectious agents while in the tomb?
- Why would an exceptionally dry environment be responsible for preserving materials in the tomb over a long period of time?

Stage 2:
THE CURSE OF THE MUMMY

Not having the benefit of training in microbiology, the scientists investigate the burial chamber of Hotsitotsi III without using any protective gear.

The mummy is found in an elaborate case filled with personal items. The mummy is left undisturbed in its case, to be analyzed later by X ray and other techniques. The small items in the case are removed, tagged, and put into plastic containers.

Around the mummy case are numerous carved wooden figurines nearly two feet tall. The Egyptian official who is with the scientists explains that these are ushabtis, who will be servants to the pharaoh in the afterlife. Several ushabtis are also packaged for further study.

The tools and material the ushabtis will need in the afterlife are also around the mummy case. An example is a bag full of wool, in excellent condition. The official explained that the ancient Egyptians believed in the ram-god, Khnum. The pharaohs were allowed to wear woolen garments as a symbol of their status as living gods, on a par with Khnum. This wool was to be used by ushabtis to make woolen clothes for the pharaoh in the afterlife.

Schmart grabs some of the wool and puts it into plastic bags. He notes that it is dry "like dry Christmas tree needles" and it pricks his hands. This is particularly painful at the site of a mosquito bite on the back of his right hand. He had scratched the bite and removed the skin from its top.

Schmart points out the wool to Dr. Noseitall, who takes a magnifying lens (like those that jewelers use) and peers into the bag of wool for several minutes. He determines that the fibers are desiccated and brittle, but similar to modern wool.

The scientists collect some mummified cats that are sitting on a shelf above the mummy case. They also scrape dust from the walls and floor of the chamber into plastic bags.

They work in the chamber for a week before packing their specimens and the mummy case and going to Cairo. The University in Cairo is providing equipment and space for the team to analyze what they have found.

Questions to ponder:

- Based on the information you are given, and assuming that there are infectious agents in the tomb, how might these agents infect the investigators?
- Animals carry diseases that can be transmitted to humans. What is the special name for such a disease? What are some examples? Can cats transmit disease to humans? Can sheep?
- What is meant by the term "reservoir"?

Stage 3:
THE CURSE OF THE MUMMY

By the time the team reaches Cairo, Dr. Noseitall has respiratory symptoms and feels ill, so the team does no further work on any material from the tomb.

Schmart (whose first name is Notso) has developed a vesicle on the back of his right hand, which has broken down to form a large ulcer with a black center.

Three days after getting to Cairo, Noseitall develops severe respiratory distress and fever. Shortly after being taken to Cairo General Hospital, he dies of a fatal septicemia.

An autopsy is done on the doctor. His organs are found to be filled with pools of blackish blood. Specimens of blood and tissue are taken for microscopic examination and culturing.

All the autopsy specimens contain gram-positive bacilli. When the organism is grown in the laboratory, it produces endospores.

Material from Schmart's lesion is also examined and shows gram-positive bacilli also. When these organisms are grown in the laboratory, they too produce endospores. Other tests done on the organisms taken from the two researchers show that they were both infected with the same type of bacteria.

Questions to ponder:

- What is septicemia?
- What are the basic shapes of bacteria and the names given to each shape? What is an endospore, and why is it of particular importance to this case?
- Bacteria are often classified as gram-negative or gram-positive. How does one carry out the gram stain? What are the results of staining gram-positive organisms and gram-negative organisms? What are the molecular differences between the two types of bacteria that result in different reactions to the gram-staining procedure?
- Determine:
  - the disease that killed Dr. Noseitall.
  - the disease that Dr. Schmart has.
  - the source of the infection of each scientist.
  - what very simple precautions the scientists might have taken to prevent these infections.
- Relate each conclusion given above to the critical information given in the narrative of the problem. In other words, tell why you came to each conclusion.

Topics Introduced by the Problem:
THE CURSE OF THE MUMMY

. . . .

Basic Background Topics in Biology
(Your textbook can be consulted for information on these.)
- Prokaryotic organisms and prokaryotic cell structure
- Bacterial shapes
- The bacterial cell wall/the gram stain and its meaning
- Bacterial endospores

Advanced Topics
(You may want to consult additional resources for information on these.)
- Resistant forms produced by microorganisms that can withstand heat, drying, and other adverse conditions
- Modes of transmission of infectious agents
- Portals of entry for infectious agents
- Protection from microbial pathogens
- Necessity of water to the vegetative forms of microbial agents
- Zoonoses
- The reservoir of a microorganism
- Septicemia
- The bacterial endospore and its unique qualities
- Anthrax and the organism that causes the disease

Additional Resources for Researching Problem Content:
THE CURSE OF THE MUMMY

Articles and books:

Hamann, B. 1994. *Disease: Identification, Prevention, and Control.* St. Louis: Mosby.

Murray, P. R., G. S. Kobayashi, M. A. Pfaller, and K. S. Rosenthal. 1994. *Medical Microbiology.* St. Louis: Mosby.

Prescott, L. M., J. P. Harley, and D. A. Klein. 1996. *Microbiology.* Dubuque, IA: W. C. Brown.

Stage 1:
THE GERITOL SOLUTION

Deborah E. Allen

John H. Martin (the former director of the Moss Landing Marine Laboratories) thought the potential problem of global warming could be addressed by dumping iron into the ocean waters off Antarctica or in other ocean areas. He and his coworkers demonstrated that the amount of chlorophyll found in ocean water samples collected (in 30-L bottles) from the Gulf of Alaska could be increased up to ninefold by the addition of iron.

When they repeated this iron-seeding experiment with water samples collected from a few hundred miles off the Antarctic coast, Martin and his colleagues found that for every unit of iron added to the sea water, the organic carbon content of that water increased by a factor of 10^4.

Martin's degree of confidence in his proposal is reflected in a remark he (half-jokingly) made during a lecture at the Woods Hole Oceanographic Institute: "Give me half a tanker of iron and I'll give you an Ice Age."

Questions to ponder:

- What is the basis for Martin's premise that seeding the ocean with iron would help combat global warming?
- What organisms found in sea water could account for the increase in chlorophyll content and biological productivity that his research group observed?
- Why did Martin choose to add iron to the water, rather than some other substance?
- Is global warming really a problem, or are environmentalists just trying to scare us into driving less and recycling more?

Stage 2:
THE GERITOL SOLUTION

Martin thought that analyses of 7,000-foot deep Antarctic ice cores provided additional support for his hypothesis. He made particular note of CO_2 and iron concentrations found 18,000 and 160,000 years ago, about the time of Earth's ice ages. This information is given in the plots shown on the following graph.

Questions to ponder:

- Do you agree with Martin that the information provided by ice core analysis supports his hypothesis?
- Why did the iron content in the ice fluctuate? Why is it lower in some areas of the ocean than others?
- Are there other likely explanations for why 20 percent of the world's oceans (including the Southern Ocean) are phytoplankton-poor?

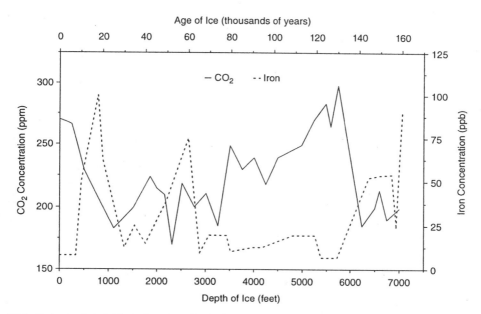

(Note that scientists think the iron found in the ice cores came from dust that settled out of the atmosphere. The CO_2 came from air trapped in bubbles in the ice.)

Stage 3:
THE GERITOL SOLUTION

You are particularly interested in Martin's ideas because you have just agreed to serve on a panel of the National Science Foundation whose task is to review grant proposals and assess their funding priority. One of the proposals you must intensively review is by a team of scientists who would like to test John Martin's iron-seeding hypothesis for the third time in open ocean waters.

You read over the proposal carefully, and then begin to consider, among other factors, whether this third attempt at iron seeding on a large scale would have the same effects reported for the previous ones.

Questions to ponder:

• Would you recommend funding of this project? What factors would you consider in making your decision?
• Are there other (better?) solutions for this problem?
• How much iron would have to be added to the ocean to reduce the impact of each year's excess CO_2 emissions? What factors do you need to consider to make this estimate?

Topics Introduced by the Problem:
THE GERITOL SOLUTION

Basic Background Topics

(Your textbook can be consulted for information on these.)
* Phytoplankton
* Photosynthesis—light-dependent and light-independent reactions
* Marine ecosystems
* Global carbon and energy cycles
* Global climate changes

More Advanced Topics

(You may need to consult additional resources for information on these.)
* Oceanic carbon cycles
* The biological pump
* A more in-depth look at marine ecosystems
* A more in-depth look at the life histories and nutritional requirements of marine phytoplankton

Additional Resources for Researching Problem Content:
THE GERITOL SOLUTION

....

Articles and book chapters:

Barnes, R., and R. Hughes. 1988. *An Introduction to Marine Ecology.* Oxford: Blackwell.

Vitousek, P., P. R. Erlich, A. H. Erlich, and P. A. Matson. 1986. "Human Appropriation of the Products of Photosynthesis." *Bioscience* 36:368–373.

White, R. M. 1990. "The Great Climate Debate." *Scientific American* July:36–43.

Electronic resources:

"Carbon Dioxide." *About Geoscience.*
<http://www.geo.nsf.gov/~develop/geo/adgeo/geofield/atmfield/co2.htm>

Farabee, M. J. *Photosynthesis.*
<http://www.emc.maricopa.edu/bio/bio181/BIOBK/BioBookPS.html>

"Oceans, Carbon, and Climate Change. An Introduction to the Joint Global Ocean Flux Study." *About Geoscience.*
<http://www.geo.nsf.gov/~develop/geo/adgeo/geofield/ocefield/co2.htm>

United Nations Environmental Program. "Oceans and the Carbon Cycle."
Climate Change Fact Sheets. <http://www.unep.ch/iucc/fs021.html>

KRYPTONITE IN HIS POCKET?

Richard S. Donham

For many years professional competitive bicycle racing in Europe has surpassed that in the United States. Similarly, while top professional cyclists in Europe achieved the status of sports heros, in the United States the best cyclists were hardly recognized. This changed dramatically in the late 1980s with the emergence of Greg LeMond as an authentic cycling champion.

Both as an amateur and a professional cyclist, LeMond established himself by winning numerous events, among them the Coors Classic in 1981, the World Championship in 1983 and 1985, and then, in 1986, the Tour de France, arguably the most demanding and certainly the most prestigious bicycle race in the world. Basically, the Tour is a multistage circuit of France, and it is the mountain stages in the Pyrenees and the Alps that make this one of the most grueling physical challenges in any sport. These include many steep climbs up narrow roads where the best athletes can "blow up" followed by terrifying descents around hairpin turns at speeds approaching 70 miles per hour. Because LeMond seemed to thrive in these conditions, his competitors sometimes referred to him as "LeMonster."

However, just as LeMond was reaching the top of his form, he was nearly killed in 1987 while hunting in northern California with his brother-in-law. As a result of a horrible accident, LeMond ended up on his back with a massive shotgun wound and pellets in his right lung, kidneys, diaphragm, and intestine. Despite extensive surgery, nearly 30 pellets remain, including two in the pericardium. Fortunately, none of these proved life threatening.

Probably LeMond's greatest achievement is his amazing recovery from this tragedy. Within two months he was back on his bicycle, beginning the long, deliberate road back to competition. Finally, he again won the Tour in 1989, overcoming a 50-second deficit in the short last stage in what is regarded as one of the most exciting finishes in the history of the sport. He repeated this in 1990, won the World Championships again in 1989, and the Tour DuPont in 1990. He was *Sports Illustrated*'s Sportsman of the Year in 1989 (see *Sports Illustrated* 71[27]: 54–75, 1989/90), was named Athlete of the Year by ABC's *Wide World of Sports*, and received the Jesse Owens Award.

After 1990, things began to go awry. LeMond's performance plummeted. He either performed poorly or repeatedly had to drop out of races because of muscle cramps. In December 1994 he announced his retirement from competition, saying he had been diagnosed as suffering

from mitochondrial myopathy, a form
of muscular dystrophy. This diagnosis has been greeted with skepticism
by some. His critics charged that
LeMond was just trying to use this
rare disease as an excuse for his per-
formances and that success had made
him lazy. *VeloNews* ran a cartoon
that gave the reason as "LeMond
can't get his pudgy arms up in the air
anymore!"

Questions to ponder:

- Assuming LeMond wants to
  respond to his critics, what is there
  about this disease that might affect
  his ability to perform at a competitive level?
- On what basis might his critics legitimately challenge his response?
- How does a muscle cell of a person
  with mitochondrial myopathy differ
  from that of an unaffected individual?
- If Greg LeMond wanted to base part
  of his argument on genetic evidence, what kind of evidence could
  he most easily provide?

Topics Introduced by the Problem:
KRYPTONITE IN HIS POCKET?

• •

Basic Background Topics

(Your textbook can be consulted for information on these.)
- Cell organization and division of labor
- Cellular energy transformation (in particular, respiratory pathways)
- Extranuclear DNA

More Advanced Topics

(You may want to consult additional resources for information on these.)
- The pathophysiology of mitochondrial myopathy
- The structure and function of muscle fibers

Additional Resources for Researching Problem Content:
KRYPTONITE IN HIS POCKET?

Articles:

Johns, D. R. 1995. "Mitochondrial DNA and Disease." *New England Journal of Medicine* 333:638–644.

Wallace, D. C. 1997. "Mitochondrial DNA in Aging and Disease." *Scientific American* 277:40–47.

Electronic resources:

Harding, A. "Mitochondrial Myopathies."
<http://www.sonnet.co.uk/rupert/mitomyop.htm>

United Mitochondrial Disease Foundation. "About Mitochondria and Disease."
<http://biochemgen.ucsd.edu/UMDF/AboutMitoDisease.htm>

Stage 1:
JIMMY HARRIS
Deborah E. Allen

....

Jimmy Harris, an eight-year-old African American boy, is brought into the emergency room of Christiana Hospital by his parents. He appears pale and jaundiced and has been passing dark red urine. His mother reports that Jimmy's friends confessed that they had dared him to swallow some mothballs, but hadn't stuck around long enough to know just how many he may have swallowed (or if he had swallowed any at all).

The results of blood tests and urinalysis show elevated levels of methemoglobin in the blood, presence of "Heinz bodies" in some of Jimmy's red blood cells (and of pieces missing from the borders of many others), and hemoglobinemia and hemoglobinuria. Subsequent assays of red cell hemolysates reveal a 95% reduction in activity of the enzyme glucose-6-phosphate dehydrogenase (G6PD).

Dr. Mancado, the admitting physician, arrives at a preliminary diagnosis.

Questions to ponder:

- Why do you think the lab has reported Jimmy's enzyme assay results as a reduction in activity?
- What would be your diagnosis of Jimmy's condition?
- Outline the series of steps by which ingestion of mothballs could have led to Jimmy's symptoms.
- Would individuals with close to 100% of normal G6PD activity respond the same way as Jimmy did to ingestion of mothballs?

Stage 2:
JIMMY HARRIS

Dr. Mancado decides to brush up on the particulars of Jimmy's genetic disorder the night before Jimmy's release from the hospital so that he is sure to offer his parents the right advice about Jimmy and his siblings (two sisters and a brother). The medical textbook he consults brings back memories of a college biology class—there on page 303 are the graphs (frequency distributions) that his professor put on the board when introducing the subject of X-inactivation. The graphs (see below) show data from an early study on a population with a high incidence of the disorder from which Jimmy suffers.

"I never did get the exam questions right on that second graph on the incidence in females," he muses. His last thoughts before falling asleep

(his book never fails to act as a potent sleeping aid) are about what he'll say to Jimmy's parents the next day.

Questions to ponder:

* What inheritance pattern would best fit the data from the graphs on page 303 of Dr. Mancado's book?
* Why is there such a variable penetrance in heterozygous females?
* What precautions should Dr. Mancado advise the Harrises to take with Jimmy and his male and female siblings?
* The disorder from which Jimmy suffers usually has a higher incidence in geographical areas where malaria is endemic. How could you explain this phenomenon?

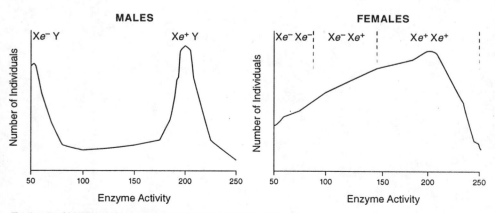

Textbook of Medicine: Hematologic and Hematopoietic Diseases

Topics Introduced by the Problem:
JIMMY HARRIS

Basic Background Topics
(Your textbook can be consulted for information on these.)
- Structure and function of proteins
- Structure and function of enzymes (which are proteins)
- Hemoglobin (another protein) and oxygen transport
- Oxidation-reduction reactions
- Coenzymes and biological reactions
- Glycolysis
- Structure of the typical animal cell
- Mendelian genetics—basic principles of inheritance, plus inheritance of sex-linked traits
- How genetic sex is determined (X-Y system)

More Advanced Topics
(You may want to consult additional resources for information on these.)
- Specific information on Jimmy's disorder (once you diagnose it)—biochemistry, clinical features, and genetics
- Malaria
- Structure and function of red blood cells

Additional Resources for Researching Problem Content:
JIMMY HARRIS

Articles and books:

Champe, P. C., and R. A. Harvey. 1994. *Biochemistry.* Philadelphia, PA: J.P. Lippincott.

Electronic resources:

Agency for Toxic Substances and Disease Registry. *Public Health Statement.* <http://atsdr1.atsdr.cdc.gov:8080/>

King, M. Terre Haute Center for Medical Education. *Medical Biochemistry Homepage.* <http://web.indstate.edu/thcme/mwking/home.html>

National Center for Biotechnology Information. *Online Mendelian Inheritance in Man (OMIM) Home Page.* <http://www3.ncbi.nlm.nih.gov/Omim/>

WHEN TWINS MARRY TWINS

Deborah E. Allen

Sally Thompson meets Harry Branaugh in her junior year at a small liberal arts college in Massachusetts. It's a case of love at first sight. In the spring of their senior year, they both have been lucky enough to find jobs in the Boston area, so they plan to get married in the June following graduation.

At their wedding rehearsal dinner, Sally's twin sister Emma meets Harry's twin brother Ken for the first time. It's a case of love at first sight. As Sally and Harry have their first serious argument about who should have told whom about having a twin (and exactly when), Emma and Ken make plans for the evening that don't include the rest of the family. Three months later, they also decide to get married.

The couples keep in touch, and three years later Sally and Emma are delighted to discover that they are both expecting (twins?). Emma's due date is in October, and Sally's in December. On December 12th, seventeen hours into labor, Sally is no longer sure she's delighted about the prospect of motherhood, and begins to worry about the child she's about to deliver.

"Why didn't you think of it sooner," she says to Harry, gripping his arm rather severely. "Identical twins should never marry identical twins. Our child's going to look just like Emma and Ken's little boy." Her first impression of Kenneth Jr., she recalls, was that he had the sort of face that only a mother and father could love.

Two hours later, Sally is scared to take a look as the obstetrical nurse puts her first child into her arms.

Questions to ponder:

- Will their child look just like his or her "double cousin," Ken Jr.? Why or why not?
- Assuming that Sally is right and the children will look identical, will they also have similar personalities, behavior, and attitudes?
- What is the maximum percent of the two childrens' genetic composition that could consist of identical genes (allelic versions)? The minimum percentage?

Topics Introduced by the Problem:
WHEN TWINS MARRY TWINS

Basic Background Topics

(Your textbook can be consulted for information on these.)
- Cell division
- The early stages in the formation and development of an embryo
- Mechanisms of genetic inheritance
- Twins

Additional Resources for Researching Problem Content:
WHEN TWINS MARRY TWINS

. . . .

Electronic resources

Angell, G. *TWINSource Home Page.*
<http://www.modcult.brown.edu/students/angell/TWINSource.html>

Berkowitz, Ari. "Our genes, ourselves?" (Article from *BioScience* 46:42–51,1996; posted with permission). *Serendip.*
<http://serendip.brynmawr.edu/gen_beh/Berkowitz.html>

University of Minnesota–Twin Cities, Department of Psychology. *MinnesotaTwin FamilyStudy.*
<http://cla.umn.edu/psych/psylabs/mtfs/tfsindex.htm>

HOW DOES *SILEX STROBILUS* REPLICATE?

Deborah E. Allen

....

As the sole Terran member of the Interplanetary Bios ID Team, Montana has been sent to the planet Linnaeus IV to investigate the mode of reproduction of a newly discovered organism, *Silex strobilus.* The first team members to arrive on Linnaeus IV had thought it was uninhabited, until one day Geieaque!, a geologist from the planet Eldore, noticed that one of the tiny "rocks" (or was it a petrified pine cone?) started to glow after he stepped on it.

After several Linnaean-years of investigation, the team has determined that one of the large molecules found in relative abundance in the rubbery core of the *Silex* has a number of properties that suggest it might serve as an informational macromolecule, the role that DNA plays in most organisms on Montana's home planet. Although primarily silicon-based, the molecule *is* a polymer consisting of two parallel chains. The chains are stabilized by bonds between atoms in the subunits that are joined to the two chains in opposite but equivalent positions.

Montana has been assigned the task of finding out if the molecule self-replicates, and if so, whether it does so in a manner similar to that of Terran DNA. Fortunately, she finds out that the mode of nutrition of *Silex*

strobilus enables her to use an experimental method she first read about many light years ago in her biology class on Earth.

She grows the *Silex* in a medium containing heavy nitrogen (^{15}N) for several generations (or replications, if this is in fact what the molecule does). By the sixth replication, extraction of the putative informational macromolecule from the organisms she has grown, followed by equilibrium density-gradient centrifugation of the molecule in solutions of CsCl, convinces Montana that all of the ^{14}N (light nitrogen) normally present in the molecule had been almost completely replaced by ^{15}N that the organisms had picked up from the growth medium.

She then divides the *Silex* into two groups; she puts Group I back into a medium containing ^{14}N (i.e., no longer having any ^{15}N) for one replication, and Group II back in an ^{14}N medium for two replications. At the end of each group's experimental period, Montana extracts the molecule from the *Silex* and centrifuges it. In the case of Group II, she also separates the two chains of the molecules before centrifuging after the second replication in order to compare the result with that of unseparated chains. She has perfected an analytical technique that enables her to determine the resulting position (top to bottom) of the molecule in each of her centrifuge tubes. She arbitrarily assigns a position of 0 tube units to the surface of the CsCl solution at the top of the tube, and a position of 20 tube units to the bottom of the tube.

Questions to ponder:

- On what basis was Montana convinced that "all of the ^{14}N (light nitrogen) normally present in the molecule had been replaced by ^{15}N that the organisms had picked up

from the growth medium"?
- Using Terran terminology, what mode of replication does *Silex* use? If this experiment had been performed on Earth using the bacterium *E. coli*, what would you predict would be the outcome?
- Is *Silex strobilus* "living" if it isn't carbon-based? Is it enough to have a large molecule that self-replicates to be considered a living organism?
- In what ways could a silicon-based polymer be similar (in structure and function) to DNA? In what ways might it have to be different?

Date: 04-01-32   Run #: 2   Montana W.

EXPERIMENTAL CONDITION	MOLECULE POSITION (TUBE UNITS)
Before transfer to ^{15}N (Groups I & II)	3
Before transfer back to ^{14}N (Groups I & II)	17
One generation after transfer Back to ^{14}N (Group I)	3, 17 (two bands)
Two generations after transfer Back to ^{14}N (Group II)	3, 17 ←
Separated strands two Generations after transfer Back to ^{14}N (Group II)	3, 17 ←

These second-generation bands look narrower than the first generation — I'll have to follow up on this.

Topics Introduced by the Problem:
HOW DOES *SILEX STROBILUS* REPLICATE?

Basic Background Topics

(Your textbook can be consulted for information on these.)
- Basic conceptual models for DNA replication
- The Meselson-Stahl experiment on replication
- The structure of DNA
- The fundamental characteristics of life

More Advanced Topics

(You may want to consult additional resources for information on these.)
- Equilibrium density-gradient centrifugation
- Experimental evidence for modes of DNA replication in eukaryotic cells

Additional Resources for Researching Problem Content:
HOW DOES *SILEX STROBILUS* REPLICATE?

••

Articles and books:

Alberts, B., D. Bray, A. Johnson, J. Lewis, M. Raff, K. Roberts, and P. Walter. 1998. *Essential Cell Biology.* New York: Garland.

Felsenfeld, G. 1985. "DNA." *Scientific American* 253 (October):58–67.

Meselson, M., and F. Stahl. 1958. "The Replication of DNA in *E. coli.*" *Proceedings of the National Academy of Sciences USA* 44:671–682.

Taylor, J. H. 1958. "The Duplication of Chromosomes." *Scientific American* 198:36–42. (Taylor describes his and colleagues' experiments in which chromosomes in onion root cells are labeled with ^{3}H and followed through successive replications/cell division.)

Stage 1:

TO BE TESTED OR NOT TO BE TESTED

Linda K. Dion

"It is but sorrow to be wise when wisdom profits not."

(The blind seer Tiresias to Oedipus in Sophocle's Oedipus the King, as recounted by Nancy Wexler, president of the Hereditary Disease Foundation, in the *FASEB Journal*, Vol. 6, 1992)

"So if we can conquer outer space, we should be able to conquer inner space too. And that's the frontier of the brain, the central nervous system, and all the afflictions of the body that destroy so many lives and rob our country of so much potential....Now we know the nerves of the spinal cord can regenerate. We are on our way to helping millions of people around the world, millions of people like me, up and out of these wheelchairs." (Christopher Reeves, 1996 Democratic National Convention, August 26, 1996)

These words of Christopher Reeves, spoken last night on television from the 1996 Democratic National Convention, are still ringing in your ears. You remember them as you sit at your first meeting of the support group of the Hereditary Disease Foundation. You came to this meeting because advertisements said that genetic counselors would explain the basis of genetic testing and the pitfalls of conducting such tests. At the age of 26, you are in your first

trimester of pregnancy and you notice that another young woman across from you is also pregnant. She has come with an older woman, presumably her mother, whose strange twitches and contortions are all too familiar.

Your father, who has displayed these involuntary movements for the past year, was recently diagnosed with Huntington's disease. Huntington's disease is an inherited neurodegenerative illness that kills by methodically destroying its victims' brain cells, causing choreic movements, slurred speech, clumsiness, and eventually dementia. Although its victims are not paralyzed as Christopher Reeves is, you hope that his plea for research on the central nervous system will lead to work that also helps Huntington's sufferers.

A genetic counselor calls the meeting to order and explains what is known about the cause of Huntington's. In 1993 researchers at Boston's Massachusetts General Hospital located the gene for

Huntington's—it is on chromosome 4. The disease-causing mutation in this gene is an expression of the triplet CAG. Instead of this triplet being present only about 18 to 19 times in the gene, as it is in unaffected people, it is repeated an average of 44 to 45 times (maybe as many as 120 times) in people with Huntington's.

Questions to ponder:

- Why would this mutation make any difference in a person's phenotype?
- What amino acid would be repeated in any protein product from this gene?
- Suggest how such a mutation might occur.

Stage 2:
TO BE TESTED OR NOT TO BE TESTED

The genetic counselor goes on to explain that the product of the Huntington's gene is a protein called huntingtin whose function is not yet known. Not only is its function unknown, but it has not yet been explained why a mutated form of huntingtin causes brain neurons to sicken and die. However, she continues, research groups are getting closer to an answer. For example, researchers at Duke University just identified a protein that interacts with huntingtin—the enzyme, glyderaldehyde-3-phosphate dehydrogenase (GAPDH).

A preliminary hypothesis, according to the counselor, is that the mutant huntingtin protein binds to the enzyme in a way that inhibits its function.

You timidly raise your hand and interrupt the counselor with a question. "Is huntingtin found throughout the body, and if so, why does the mutant form affect only cells of the brain?"

"Yes," she answers, "it's found throughout the body, but it's our brain cells that rely almost exclusively on glucose for their energy."

You mull this over as the counselor responds to someone else's question, but you are distracted with worry over your own predicament. Although currently only 30,000 people in the United States have Huntington's disease, your father is one of them (he was diagnosed when he was 45), and you are fearful that you could have passed the gene on to your unborn child. On the other hand, you don't have the disease, so you wonder if you should even be concerned about it.

Questions to ponder:

- Why would the binding of mutant huntingtin to the enzyme GAPDH have an effect on a cell?
- Why is the counselor's explanation that brain cells rely principally on glucose for energy relevant to your question about why these cells are the ones that are most affected by mutant huntingtin? What is the missing (final) step that would make her explanation complete?
- Should you be concerned about getting Huntington's disease or about passing the gene on to your child? Why or why not?
- Considering that afflicted people inevitably die from this disease, why do 30,000 people have it? Why has the gene not been eliminated from the population?

Stage 3:
TO BE TESTED OR NOT TO BE TESTED

You snap out of your reverie to hear the counselor explaining genetic testing. For an adult, a simple blood sample is drawn; for an unborn child, amniocentesis or chorionic villus sampling of the placenta will yield a sample of fetal cells. From either type of cell, a specimen of pure DNA is extracted. This specimen is amplified to the desired quantity, and then cut with enzymes into fragments of different sizes. This is possible because each of the enzymes recognizes only certain nucleotide sequences, so each cuts the DNA at different places along its length.

These DNA fragments are then separated according to size, then stained so their pattern of arrangement can be seen. For the Huntington's gene, people of different genotypes (Hh and hh) produce different fragment patterns (because the enzymes have cut their DNA into fragments of different lengths). Because these differences exist, the human population is said to be polymorphic for the fragments. The different patterns are called RFLP's (short for restriction fragment length polymorphisms), and they serve as markers for either the Hh or the hh genotype.

"And so," concludes the counselor, "each one of you has the right to choose whether to be tested or not. Do not let anyone pressure you into doing it—the choice is yours and yours alone."

You feel overwhelmed by the conflicting opinions you have about testing and know that this will not be an easy decision to make.

Questions to ponder:

- How is it possible for a DNA cutting enzyme to recognize some DNA nucleotide sequences but not others?
- Why would a person who has Huntington's disease have a different RFLP pattern than someone who is unaffected?
- Suppose the test came out positive and you find that you are heterozygous for Huntington's disease. What are the benefits and pitfalls of having this knowledge? How would this be different from learning that you are heterozygous for cystic fibrosis, another genetic disease?

Topics Introduced by the Problem:
TO BE TESTED OR NOT TO BE TESTED

Basic Background Topics
(Your textbook can be consulted for information on these.)
- Transcription
- Translation
- Glycolysis
- Cellular respiration
- Mutations
- Protein structure
- Enzymes
- Amniocentesis and chorionic villus sampling
- An introduction to methods for genetic testing

More Advanced Topics
(You may want to consult additional resources for information on these.)
- Huntington's disease
- Neurodegenerative diseases
- Trinucleotide repeats
- Fetal tissue transplantation
- The value of predictive testing and genetic counseling
- RFLP analysis

Additional Resources for Researching Problem Content:
TO BE TESTED OR NOT TO BE TESTED

Articles:
Barinaga, M. 1996. "An Intriguing New Lead on Huntington's Disease." *Science* 271:1233–1234.

Wexler, N. S. 1992. "The Tiresias Complex: Huntington's Disease as a Paradigm of Testing for Late-Onset Disorders." *FASEB Journal* 6:2820–2825.

Wiggins, S. P. et al. 1992. "The Psychological Consequences of Predictive Testing for Huntington's Disease." *New England Journal of Medicine* 327:1401–1405.

Electronic resources:
Collins, D. University of Kansas Medical Center. "Genetics of Huntington's Disease." *Caring for People with Huntington's Disease Page.* <http://www.kumc.edu/hospital/huntingtons/genetics.html>

Hereditary Disease Foundation Homepage. <http://www.hdfoundation.org/index.html> (For general information about Huntington's disease; and more specific information about the work of Nancy Wexler, the president of the foundation; the search for Huntington's gene; and the triplet repeats of Huntington's.)

Stage 1:
ANNA OR ANASTASIA?

Deborah E. Allen

....

One summer's night in 1918, a fusillade of shots rang out from a room deep within the Ipatiev House near Ekaterinburg (now Sverdlovsk), a prosperous mining town in the Urals. After a year's imprisonment, Tsar Nicholas II, the last emperor of Russia, his wife Tsarina Alexandra and their five children, along with three family servants and the family physician, were executed by firing squad.

When pieced together, a secret account left by the chief executioner (a Bolshevik named Yakov Yurovsky) plus a seven-volume dossier written by a White Russian monarchist who officially investigated the execution in 1918 to 1919, paint a vivid picture of how and where the victims were buried. According to these sources, the bodies were transported by truck to a forest clearing several hundred yards from the Ekaterinburg-Perm railway line, stripped of their clothing, in at least some cases badly burned (most likely by sulfuric acid), and then unceremoniously dumped in a relatively small and shallow pit. According to Yurovksy's account, this was not the victims' first burial site—two days after the murder, the bodies were retrieved from the mine shaft down which they were initially deposited on the day of execution, and then reburied out of fear that they might be discovered by the approaching White Army.

Although the grave was located in 1979 (by a geologist and a film maker, who removed, then returned three of the skulls), the bodies were not exhumed until 1991. The excavators used methods later referred to as "rude violations of archeological and forensic norms" by at least one expert, a Professor Krukov of Moscow (as quoted by Robert Massie in *The New Yorker* August 21 & 28, 1995). Permission to exhume was granted by Boris Yeltsin, who had just been elected President of Russia after dissolution of the Central Committee of the Communist Party.

The exhumed skeletons told a striking tale of further brutality—of repeated bayoneting, faces smashed by rifle butts, and broken jaws. A month after excavation of the 75-year-old burial site, Dr. Sergei Abramov, a forensics expert from the Russian Ministry of Health, meticulously began to piece together nearly 1000 bones and bone fragments to assemble nine skeletons (rather than the expected number of eleven). With the aid of a computer, the skulls were matched with photographs of the Romanovs. As a result of this facial reconstruction and additional forensic evidence (odontology, sexing, and age determination), Abramov and his team were convinced that they had identified the remains of Nicholas II and Alexandra, their physician and three servants, and three of their five children (the Grand Duchesses Olga, Tatiana, and Anastasia).

An equally dramatic aspect of these findings was what they implied —that the bones of (the former) Tsarevich Alexis (son and heir of Nicholas II) and his older sister Marie (as Abramov contended) were missing from amongst the exhumed remains. Was it actually possible that these two had survived the execution? Or, did this confirm Yakov Yurovsky's chilling story that the bodies of the Tsarevich and an unidentified (by him) female were not buried along with the other victims, but burned and the remains scattered near the burial site?

These findings were made public in a two-day conference held in Sverdlovsk (Ekaterinburg) and attended by Dr. William Maples, a forensic anthropologist from the University of Florida (and director of the C. A. Pound Human Identification Laboratory in Gainsville). Just prior to the conference, Maples had (unbeknownst to Abramov) put together an impressive team of experts and made a pitch to local authorities for permission to conduct further analyses of the exhumed remains. His team successfully displaced Abramov's Moscow-based team and another impressive U.S. group (assembled by the FBI and the Armed Forces Institute of Pathology) from involvement in this subsequent analysis.

Upon arrival in Sverdlovsk, Maples and his colleagues found the exhumed skeletons lying unprotected (and vulnerable to temperature fluctuations and sunlight exposure) on separate metal tables in a room in the local morgue. Some of the long bones had been severed into sections, a fact

that would impede accurate measurement of length, an important aspect of the identification process. Despite these constraints, the Maples team's work confirmed many of the earlier findings of the Abramov team, with one important difference: according to Maples, none of the skeletons were young enough to be the bones of Anastasia. He asserted that the third daughter present in the grave was actually Marie, not the Tsar's youngest daughter, Anastasia.

Dr. Abramov, however, could not be swayed from his original conclusion that the bones in question belonged to Anastasia. He, after all, had based his analysis on measurement of the femur before it had been sectioned, and immediately after it had been laid out and assembled by him as part of the skeleton whose identity was disputed. (Was his insistence on the accuracy of his earlier finding somewhat colored by his feelings about being summarily displaced when this new forensic team appeared in Sverdlovsk?). Maples's identification was based on analysis of the radius and ulna, which may well have been moved by the time he got to them, mislaid to rest among the unprotected bones whose ownership was in dispute.

Maples in turn found fault with the techniques Abramov used to assemble and subsequently identify the various skull fragments, particularly in cases in which some of the identification landmarks were missing. He also highlighted his evidence from study of the vertebrae—all of those found in the grave had completely fused rings on their body (annular epiphyses), a characteristic of adult skeletons. He contended that signs of immaturity (instances of partial fusion) is an expected finding for the vertebrae of a 17-year-old, Anastasia's age at the time of the execution. Anastasia therefore could not have been present in the mass grave.

Both scientists however, agreed on the essential point—these *were* the remains of the Romanovs.

The controversy over who was the missing daughter seemed to be resolved when two independent teams (one headed by the British researcher, Peter Gill, director of Biology Research at the British Home Office Forensic Science Service in Aldermaston) performed DNA analysis on the skeletal remains. Because DNA technology was not then available in Russia, pieces of the bones from the nine skeletons were transported (hand-carried in an airline travel bag) to England for the DNA fingerprinting analysis by Gill.

As a prelude to the analysis performed under Gill's supervision, the old and badly deteriorated bones were ground to a fine powder, and subjected to a combination of processes that separated a minuscule quantity of DNA from other chemicals present in the bones. The even smaller quantity of nuclear DNA included in the DNA extract was used to establish the sex of the victims, through a procedure based on amplification (using PCR, or polymerase chain reaction) of a portion of the amelogenin gene, which is homologous to the X and Y chromosomes. The nuclear DNA also proved suitable for use in establishing (through the fingerprinting process) that the male presumed to be Nicholas was the father of the three youngest women whose bodies had also been in the shallow grave. This latter analysis of the nuclear DNA was not performed using the (then) more standard method of RFLP (restriction fragment length polymor-

phism) analysis, but by a newer gene amplification analysis based on examination of inheritance patterns of chromosomal short tandem repeat (STR) loci (also called microsatellites). For other essential features of the identification process, the team turned to the mitochondrial (mt) DNA present in the extracted samples. Hypervariable mtDNA regions (from 634 to 782 base pairs in length, and from both DNA strands) were sequenced for each of the victims, and the results compared.

A problem of a nature quite different from that of the difficult task of extracting DNA from old bones soon surfaced. Would it be possible to find contemporary relatives of the slain Russian royal family who would agree to donate blood for use in the analysis? Alexandra's nearest surviving relatives were relatively easy to locate. Of the two possibilities, a grandniece and grandnephew, it was the latter (a.k.a. Prince Philip, Duke of Edinburgh and consort of Queen Elizabeth II of England) who agreed to donate his blood in the attempt to confirm Alexandra's identity. The search for Nicholas II's closest surviving relatives proved more difficult— his nephew (Tikhon Kulikovsky) refused to cooperate out of contempt for the fact that the British refused to offer the Tsar and his family refuge after the revolution. Eventually a female descendant of Nicholas's sister the Grand Duchess Xenia, and a male descendant of Louise of Hesse-Cassel, Nicholas' maternal grandmother, agreed to be donors. The sample from the female descendant, Xenia Sfiris, was sent by diplomatic pouch from the British embassy in Athens in the form of dried blood that Mrs. Sfiris had dripped on a paper napkin after pricking her finger.

Table 1. Four of the five STR genotypes obtained from putative skeletons of the Romanov family

Skeleton*	HUMVAWA /31**	HUMTH01	HUMF13A1	HUMFES/FPS
Tsar	15,16	7, 10	7, 7	12,12
Tsarina	15,16	8, 8	3, 5	12, 13
Child	15,16	8, 10	5, 7	12, 13
Child	15,16	7, 8	5, 7	12, 13
Child	15,16	8, 10	3, 7	12, 13

*Designations of the skeletons are as reported in *Nature Genetics*.
**Allele designations for all loci are based on the number of repeat units that were found by sequencing of specific alleles.

The mtDNA analysis generated identical sequences from the putative Tsarina and her three children; the sample of blood from Prince Philip allowed for confirmation of the identification of the mother and of the sibling status of these three young females. Comparison of the mtDNA sequence of the putative Tsar with his two contemporary relatives (of unbroken maternal descent from his maternal grandmother) revealed a match, with the exception of a single nucleotide at position 16169. Subsequent analyses indicated that this mismatch at 16169 was attributable to heteroplasmy at this position in the Tsar's mtDNA, an unusual finding. Table 1 above, reproduced in part from the *Nature Genetics* article (Vol. 6, 1994) in which Gill and colleagues reported the results of their ten months of work, summarizes a major portion of the findings from the STR analysis (four of the five tetrameric loci that were analyzed) on the Romanov family remains.

In July 1993 Gill and colleagues announced their findings to the world in a press conference in London— these were *indeed* the remains of the Romanovs. In early 1994 the findings were published in the prestigious journal *Nature Genetics* (Vol. 6:130– 135). In the discussion section of this article, the authors alluded to the dispute over the identity of one of the two missing children in the following rather dispassionate manner: "Although two of the Romanov children were missing from the grave, we cannot speculate on the identity of the missing princess."

Questions to ponder:

- What is the source of DNA in bones that have been in the ground for over seventy years? What are the possibilities?
- How can a gene segment that is present on both the X and Y chromosomes be used for sex testing? What findings might indicate that an individual is male rather than female?
- Why did the British DNA fingerprinting experts need to turn to mitochondrial DNA for further analysis? Of what particular usefulness was this extranuclear DNA to the identification process? On the other hand, what were the limitations, in terms of establishing identity, to use of the mitochondrial DNA?
- Why do you think that STR analysis was used instead of the more common and well-established method of fingerprinting, RFLP analysis? What can you conclude about this analysis on the basis of the findings reported in Table 1? What additional findings (from the STR analysis)

might allow you to strengthen these conclusions?
- The forensic anthropologist Maples accused Gill and colleagues of what essentially amounts to doing sloppy work. (Was Maples's accusation in part motivated by his dismay over not being selected to direct the DNA analysis?) He criticized them for using the long bones as a source of DNA (after all, these bones might have gotten shuffled around), and asserted that the findings from analysis of the Tsar's mtDNA were more suggestive of contamination of the samples than of the genetic anomaly of heteroplasmy. (Maples's claim may have been at least in part responsible for delay by the Patriarchal [Russian Orthodox] Church in Russia in deciding whether to canonize the imperial family, and on plans for a funeral and ceremonial interment.) On what basis, if any, could Maples legitimately make a claim of sample contamination? What counter arguments could Gill have used to help resolve this dispute?

Stage 2:
ANNA OR ANASTASIA?

Other people, however, were more than willing to speculate on the identity of the missing female child of Nicholas and Alexandra. The speculation began in the years immediately after the execution of Tsar Nicholas and his family, shortly after the rescue of a young woman from near drowning in a German canal. She recovered from her suicide attempt in a public mental hospital in Berlin, referred to by the hospital staff (who assumed that she was a Russian refugee) as "Miss Unknown," since she seemed to suffer from amnesia.

In 1921 "Miss Unknown" at last spoke up and publically announced her true identity—she was Anastasia Romanov. She went on to tell a fantastic story of having been rescued by one of her captors (a kind-hearted executioner?) and smuggled across the border into Romania. She then came to Germany seeking Romanov relatives, throwing herself into the canal when she failed to make contact.

Anna Anderson, as she later came to call herself, became an international celebrity—numerous books and films went on to flesh out her story, shaping it into a fairy tale legend of a lost princess. Twelve members of the Romanov family were not so willing to believe in fairy tales, at least not this one—they issued a statement in 1928 that publicly denounced Anna Anderson. One (Anastasia's uncle, the Grand Duke of Hesse) hired a detective, who found evidence that

Anna Anderson was not Anastasia Romanov, but a Polish woman named Franzisca Schanzkowska. Schanzkowska had disappeared from her rooming house in Berlin shortly before Anna Anderson was rescued from the canal, and had been injured from an explosion in the hand grenade factory in which she worked.

Anna Anderson did have supporters, however, including members of the Romanov family (Anastasia's cousin the Princess Xenia, for example) and other members of the European aristocracy. They based their faith in her declared identity on evidence such as the following physical similarities to Anastasia: height, hair and eye color, a cauterized mole on the shoulder, a distinctive bunion on the toe, and walk and physical carriage. In addition, Anna Anderson had a number of scars, including one on her foot that had the star-shaped pattern of a bayonet, and provided convincing recollections of Romanov family members and childhood events in the letters she now wrote to her supporters.

Conflicting evidence came from people who had known Anastasia well. Gled Botkin, the son of the Romanov family physician (Eugene Botkin) slaughtered along with the family, was Anastasia's playmate in the year of captivity before the execution. Upon meeting Anna Anderson nearly ten years after he had last seen Anastasia, he was immediately convinced (by the way she looked and walked) that she was Anastasia. Anna-Anastasia's apparent recognition of pictures they had painted together as children, plus her strong reaction to those painted during their captivity in Siberia, were merely "icing on the cake." Gled Botkin's daughter and her husband, Marina and Richard Schweitzer, remain firmly convinced to this day that Anna Anderson and Anastasia were one and the same.

Further verification came from a Russian dragoon captain (whom Anastasia had visited in the hospital). He met Anna Anderson and showed her an old photograph that included another soldier that Anastasia had also met on her hospital visits. Anna referred to this other soldier as "the man with the pockets," a nickname conferred on him by the young Anastasia. However, Pierre Gilliard, Anastasia's former tutor, declared Anna Anderson a fraud after meeting with her several times.

How could a Polish peasant have convinced so many that she was a member of the Russian royal family? According to some, because she was a confidence trickster with a special aptitude for learning, much like famous courtesans of former times.

Establishing the identity of Anna Anderson moved to the courtroom when the Romanov family sought to declare all of the Tsar's children officially dead in order to resolve inheritance of the family fortune. Anna Anderson responded with a counter-suit, claiming that Anastasia was very much alive. In the absence of fingerprints or dental records for their use, the courts finally ruled in 1967 that Anna Anderson had not successfully proven her claim to be Anastasia.

Upon losing the lawsuit, Anna came to the United States and settled in Virginia where she married a retired history professor, changing her name to Anna Manahan. Anna Anderson Manahan died in 1984 after a lifelong struggle to establish her true identity as Anastasia. The DNA and forensic evidence suggesting that Anastasia may have been missing from the Romanov mass grave had come too late to bolster her claim.

Equally late was a computer-assisted analysis of identity based on matching of ears from photographs of Anna Anderson and Anastasia, performed in 1993 by a team of experts led by the forensic scientist Peter Vanezis. The analysis was based on the premise that no two ears are alike (even in identical twins), and that after about 4 months of age, the attachment of the ear to the face and the contours of its design do not change, even into old age. Photographs were blown up to emphasize the ears and referenced to the same size scale, then six different technicians performed the ear identification analysis. Photographs of the ears of Anastasia and Anna Anderson were mixed with those of five other women, and the technicians were unaware of the identity of any of the ears. The technicians consistently rated the extent of match between Anna and Anastasia's right ear (from one of Anna's photographs) to be a 4 (on a 1 to 5 scale). The match actually seemed perfect, but the alignment of the ears was slightly different in the two compared photographs, hence the downgrading of the verdict to a "good" rating rather than an "excellent." The inner portion of the left ear did receive a score of "5," an amazingly close match that Vanezis did not think could occur by chance alone. When matched to the photograph of the real Anastasia, none of the photographs of the five other women generated scores even close to these.

If this DNA and other forensic evidence, and evidence from ear lobe analysis had been available to the European courts, might Anna Anderson have been able to establish her claim?

Marina Botkin Schweitzer and her husband Richard never ceased their attempt to seek justice for Anna Anderson Manahan, even after her death. They tracked down a piece of small intestine to a hospital pathology laboratory; the hospital records indicated that the sample had been removed (for biopsy) from Anna during a 1979 surgical procedure. After some legal wrangling they received the green light to enlist Peter Gill to perform another DNA analysis. Gill's forensic laboratory compared the mtDNA from the biopsied specimen to that from Prince Philip, as well as to that from a surviving maternal grandnephew of Franzisca Schanzkowska, the missing Polish worker.

Before hearing the results of the analysis, Peter Kurth, Anna Anderson's biographer, long-time friend, and strong supporter of her claim to be Anastasia, made the following remark when interviewed on the PBS program *NOVA* (the episode was entitled "Anastasia - Dead or Alive?"): "If we find out that two or three reputable tests for DNA demonstrate that she is not related to those bones [those found in the Russian burial site], then I will say that it looks like we made a mistake."

NOVA viewers were also able to witness Peter Gill's phone call to the Schweitzers, informing them of the fact that the DNA in the biopsied sample of intestine did not match the profile that could be expected to be found from Anastasia, but did in fact show a positive match to that from the grandnephew of Franzisca

Schanzkowska (these results were corroborated independently by two other expert teams, one from Penn State University, and the other from the Armed Forces DNA Identification Laboratory).

"That is just unbelievable. That's just impossible," they responded in shock. "They [several aristocratic Anna supporters] would not have been taken in by somebody less than aristocratic origin...would never have been taken in by a Polish peasant."

Peter Kurth also was asked to comment on the outcome of the analysis in front of *NOVA*'s cameras, and he expressed a similar disbelief. "It is impossible for me to accept that Anna Anderson was Franzisca Schanzkowska. I knew her. I'm speaking not as an expert, but as a witness...as someone who knew what her manners, her gestures, her every fiber was made of. She was an absolute lady right down the line. There was nothing about her that made you think of factories and fields. Nothing."

He went on to comment, "It's not about Anna Anderson, these statistics about what chance it would be that she was this, that, or the other thing....It's not about her at all. It's about science. The tragedy of science and the dark side of science is that it doesn't take into account the authentic experience of real people....That's all I'm doing here now—is insisting on my own experience."

In early 1995, Peter Gill and scientists from the two other forensic laboratories summarized their findings on Anna Anderson's identity, along with findings from a later analysis of a hair sample purported to have come from Anna Anderson, in a two-page article in the "Correspondence" section of *Nature Genetics* (Vol. 9). Several months later, the same section of the well-regarded journal printed a letter from Richard Schweitzer. In it Schweitzer questions the origin of the tissue samples used by Gill and members of the other two fingerprinting teams, asserting that Gill himself referred to them as the "putative Anna Anderson samples" and was consistently careful to reference all of his conclusions to "the samples tested," rather than to a particular individual. He concludes his letter with the following statement: "The rational human experiences of persons who actually knew Anna Anderson, and of those who know the other evidence of her identity, are in direct conflict with the conclusion or inference drawn that this specific individual could possibly have been a Polish peasant of that era. When inferences drawn from science conflict with rational human experience, that conflict should be resolved. That is the task we are continuing."

Questions to ponder:

- As you weigh the total evidence for and against Anna Anderson being Anastasia, which do you find to be the most convincing? How do you in particular reconcile the diametrically opposed findings from the earlobe versus the DNA analysis used as evidence in establishing her identity?

- Do you agree with Schweitzer's and Kurth's implications that rational human experience and DNA analysis carry equal weight as evidence in establishing identity?

- Do you agree with Kurth that science has a dark side, running roughshod over the authentic experiences of humans? Do you think that Kurth's views about the value of science would be fundamentally different if Anna Anderson's and

Prince Philip's mtDNA had been a match?

- In an editorial in *Nature Genetics* (Vol. 8:205-206, 1994) the editor wondered why it was that Schweitzer and his supporters refused to accept the DNA results regarding Anna Anderson's identity. In this context the editor poses some broader questions to the journal's readers: "What, given such reluctance, does the scientific community have to do to convince the public that it knows what it is talking about and is accurate in its assessments?...Does this distrust extend to the general public which has been witness to many arguments for and against the use of genetic testing in courts of law?" The editor goes on to point out how important the (then) upcoming O. J. Simpson trial will be in terms of exerting a tremendous influence on the public's acceptance of these "powerful tools of justice." In your opinion, is Schweitzer's reluctance to accept the DNA results a general human propensity, or is his situation a special case?

- The DNA analyses that contributed to identification of the Romanov remains and to resolving the identity of Anna Anderson were undoubtedly of great historical interest. Why did *Nature Genetics* (a premier scientific journal to which scientists submit only their most intriguing data) publish a full article about DNA analysis of the Romanov remains, but only a "letter" in the Correspondence section about the identification of Anna Anderson? After all, aren't more people aware of the mystery surrounding Anastasia than they are about the fact that the Romanov remains were missing for nearly 75 years?

Topics Introduced by the Problem:
ANNA OR ANASTASIA?

....

Basic Background Topics in Biology

(Your textbook can be consulted for information on these.)
- Inheritance of nuclear and mitochondrial DNA
- Introduction to DNA fingerprinting techniques
- The nature of scientific investigations and analysis of evidence
- X-Y system of sex determination in mammals

More Advanced Topics (or Topics Outside the Immediate Realm of Biology)

(You may want to consult additional resources for information on these.)
- DNA fingerprinting—more specific aspects of RFLP and STR analysis, and using DNA analysis for sex testing
- Microsatellites
- Stability of DNA's structure
- Safeguards against sources of contamination in DNA analysis
- The historical events surrounding the Bolshevik Revolution and fall of the Tsar

Additional Resources for Researching Problem Content:
ANNA OR ANASTASIA?

The historical events depicted in this problem were based on accounts found in the following sources:

Barnes, M. (writer, producer and director). 1995. "Anastasia—Dead or Alive?" (Episode of *NOVA*). Boston: WGBH Educational Foundation.

Massie, R. K. 1995. "The Last Romanov Mystery." *The New Yorker Magazine* August 21 & 28:73–95.

Additional articles and books:

Editor. 1994. "Anastasia and the Tools of Justice [editorial]." *Nature Genetics* 8:205–206.

Edvotek Staff. 1997. *DNA Fingerprinting*. Mansfield, OH: Beckley Cardy Group.

Gill, P., P. L. Ivanov, C. Klimpton, R. Piercy, N. Benson, G. Tully, I. Everett, E. Hagelberg, and K. Sullivan. 1994. "Identification of the Remains of the Romanov Family by DNA Analysis." *Nature Genetics* 6:130–135.

Gill, P., C. Kimpton, R. Aliston-Greiner, K. Sullivan, M. Stoneking, T. Melton, J. Nott, S. Barritt, R. Roby, M. Holland, and V. Weedn. 1995. "Establishing the Identity of Anna Anderson Manahan [letter]." *Nature Genetics* 9:9–10.

Housman, D. E. 1995. "Clinical Implications of Basic Research: DNA on Trial— the Molecular Basis of DNA Fingerprinting." *The New England Journal of Medicine* 332:534–535.

Massie, R. K. 1995. *The Romanovs: The Final Chapter.* New York: Random House.

Micheli, M. R., and R. Bova. 1997. *Fingerprinting Methods Based on PCR.* New York: Springer-Verlag.

Tracey, M,. and R. J. Herrera. 1996. *DNA Fingerprinting*. Sudbury, MA: Jones & Bartlett Publishers.

Schweitzer, R. R. 1995. "Anastasia and Anna Anderson [letter]." *Nature Genetics* 9:345.

Smiley, B. 1996. "Fingerprinting the Dead." *Archeology* 49:66–67.

Wall, W. J. 1995. "Whose DNA is It Anyway?" *New Statesman and Society* 8:20–21.

Electronic resources

Michelle Kaske, DePaul University. "The Romanov Execution." <http://shrike.depaul.edu/`mkaske/index.html>

Stage 1:

A PROBLEM WITH PORE BEHAVIOR

Deborah E. Allen

The phone rings at 3 A.M.. Once again it's the hospital calling. One of your patients, Jill Stern, a 29-year-old that you have been treating since infancy, was admitted this morning with what seems to be pneumonia. Although the lab results aren't back yet, Jill has a recurring history of *Pseudomonas* infections. The hospital has called to tell you that she isn't responding to this latest round of antibiotic treatment (the same one you prescribed last time), and has reached a crisis stage. You have just about run through the available antibiotic arsenal, and you are concerned that if you can't intervene successfully, Jill's already scarred lung tissue will make death by respiratory failure a near certainty this time.

Although Jill's short life has consisted of a series of rebounds from severe medical crises—interspersed with a seemingly endless round of chemical, nutritional, and physical therapies aimed at her chronic symptoms (pancreatic insufficiency, chronic bronchial obstruction, and many others)—you desperately want her to pull through this latest crisis. Since you first diagnosed Jill's illness (by sweat chloride test) nearly 30 years ago, the gene that is defective has been identified, the function of the gene product has been discovered, and the connection between the defect and some of Jill's recurring symptoms has been made; you feel that the discovery of the right combination of life-sustaining therapies is just around the corner.

Questions to ponder:

- From what disease does Jill suffer?
- Is it possible for a mutation in just one gene to cause such a diverse array of symptoms? (Can you think of any other genetic disorders for which this seems to be the case?) Or should scientists still be looking for additional culprits?
- How can scientists identify defective genes? What are the general strategies used?
- Would Jill be a suitable candidate for clinical trials of any promising new therapies? Why or why not?
- Whereas people afflicted with the disease from which Jill suffers used to die in early childhood (as recently as in the 1960s), recent advances in therapeutic options have made it possible for afflicted individuals to survive to reach reproductive age and beyond. Can or should people with this disease have children?

Stage 2:

A PROBLEM WITH PORE BEHAVIOR

As you contemplate the future possibilities for your patient, a group known as the National Organization for Negative Eugenics (NONE) is meeting in secret in the executive suite of a hotel somewhere off the Washington, D.C., beltway. Members of this organization want to lobby Congress to enact legislation that will make it possible to eliminate harmful lethal recessive alleles from the human gene pool (think of the savings in health care costs alone!).

One of the members has developed a simple computer program that will enable the organization to calculate how long it would take to reduce significantly the frequency of various harmful alleles. He comes to the meeting with a request to use the organization's computer to crunch numbers that could bolster the case the organization would like to make

for, at the very least, mandatory genetic screening. To test the program, he wants to use your patient's disease as an example, assuming that the gene thought to cause it is in a state of equilibrium, that the incidence of the disease in the United States is 0.04 percent, and that 1 in 25 Americans is a carrier for the gene.

Questions to ponder:

- Will the scheme they have proposed work in the case of this disease? Why or why not?
- Should genetic screening for harmful recessive genes be mandatory? If so, should people who test positive be discouraged (prevented?) from having children?
- Is it ethically or morally right to try to eliminate harmful mutant genes from the human gene pool?

Topics Introduced by the Problem:

A PROBLEM WITH PORE BEHAVIOR

Basic Background Topics

(Your textbook can be consulted for information on these.)
- Active transport
- The role of the endomembrane system in processing proteins
- Inheritance of autosomal genes
- Genetic screening—modern methods used
- Hardy-Weinberg law and equilibrium

More Advanced Topics

(You may want to consult additional resources for information on these.)
- The genetics and pathophysiology of Jill's disease
- Modern therapies for Jill's disease
- Pros and cons of whether genetic screening should be mandatory for this disease

Additional Resources for Researching Problem Content:
A PROBLEM WITH PORE BEHAVIOR

Articles and books:

Cecil, R. L., and J. C. Bennett, editors. 1996. *Cecil Textbook of Medicine.* Philadelphia: W.B. Saunders.

Cook, A. R., and P. D. Dresser, editors. 1995. *Respiratory Diseases and Disorders Source Book (Volume 6, Health References Series).* Detroit: Omnigraphics.

Marshall, E. 1996. "The Genome Program's Conscience: A Research Program on the Ethical, Legal, and Social Implications of Genome Studies, Launched as an 'Afterthought' Is Now the World's Biggest Bioethics Program." *Science* 274:488–490.

Electronic resources:

Cedars Sinai Health System. *Neonatology on the Web.* <http://www.csmc.edu/neonatology/ref/ref.html>

Genetics and IVF Institute Home Page.<http://www.givf.com/>

National Center for Biotechnology Information. *Online Mendelian Inheritance in Man (OMIM) Home Page.*<http://www3.ncbi.nlm.nih.gov/Omim/>

Oregon Health Sciences University. *CliniWeb.* <http://www.ohsu.edu/cliniweb/search.html>

University of Iowa. *The Virtual Hospital Home Page.* <http://indy.radiology.uiowa.edu/>

Additional resources will be suggested to you if needed once you diagnose Jill's disease.

Stage 1:
MAD COWS OF KENT
Deborah E. Allen

It all began when a farmer in Kent, England, noticed his herd behaving rather strangely. His cows had begun walking with a peculiar gait, seemed to be in a constant state of apprehension, and were hypersensitive to sound and touch. A few of the animals became unmanageably aggressive (have you ever tried to milk a demented cow?).

This "mad cow disease" (as it came to be known in the popular press) began to spread all over England. Four years after the Kentish herd showed signs of the disease, the total number of reported cases had risen to 22,000.

The fact that no one knew how the disease was being spread was cause for initial concern among farmers, veterinarians, and officials and scientists at the agricultural ministry. When reports came in that the disease could jump to other species, including cats and laboratory mice that had been fed the minced brains of mad cows, public concern and anxiety reached epidemic proportions. Was eating hamburgers now a serious health hazard to the British public? Or, as *The Economist* queried in a banner headline, were cows now "Mad, Bad, and Dangerous to Eat"?

A clue to the mysterious origins of mad cow disease was actually deciphered back in 1986 when a neuropathologist examined specimens of brain tissue from the Kentish farmer's herd. He noticed that the tissue had an appearance similar to that of sheep infected with a condition known as scrapie (named after an interesting symptom—intense itching that prompts the sheep to rub themselves against fences or other handy scratching posts). It was spongy in appearance (so pockmarked with holes that it resembled Swiss cheese) and contained clusters of rod-like fibrils. Because of the resemblance between the infected cow and sheep brains, mad cow disease was given the formal scientific name of "bovine spongiform encephalopathy" (BSE).

Mad cow disease now had an official name, but its cause was still a mystery. Unlike a typical infectious disease of viral or bacterial origin, BSE did not invoke an immune response in its bovine victims; yet like infectious diseases, it could spread rapidly within an infected herd.

Questions to ponder:

- How do infectious diseases usually spread within and between herds of domesticated animals? What are the possible mechanisms? What modern agricultural practices help prevent their spread?

- After this first BSE scare, John Gummer, Britain's Minister of Agriculture, was photographed eating hamburgers with his 4-year-old daughter. Was this attempt to convince the British public that beef was safe to eat a good idea? What factors did you consider in forming this opinion?

Stage 2:
MAD COWS OF KENT

Stanley Prusiner, a neurologist at the University of California, San Francisco, came up with a controversial hypothesis for the origin of BSE, scrapie, and Gerstmann-Straussler-Sheinker syndrome (GSS), a spongiform encephalopathy that affects humans and runs in families. He postulated that these diseases are caused by prions (short for proteinaceous infectious particles), or abnormal, infectious proteins.

Some of the initial evidence for his hypothesis came from relatively simple observations. In the case of scrapie, for example, infected brain tissue in which the nucleic acids were destroyed by radiation or chemicals could still be used to infect lab mice. On the other hand, scrapie-infected tissue in which the proteins had been destroyed (using a combination of enzymes that break down proteins) was not infective. Finally, the brain tissue of animals with spongiform encephalopathy was found to have a protein (PrP) that is an abnormal variant of one normally kept in check by cellular enzymes. In spongiform encephalopathy, this aberrant form instead accumulates to form rod-like fibrils that resemble those found in infected brain tissue. In humans with GSS, Prusiner has found a mutation in the form of the gene producing the

normal version of the protein. Some scientists began to speculate that this aberrant form of the protein was actually the prion itself.

Prusiner's critics asked, however, how a protein (with no nucleic acid associated with it) could enter a cell and make enough copies of itself to create a disease in its host. They bolstered their first round of attacks on Prusiner's hypothesis by pointing out that when this abnormal cellular protein is cloned and injected into test animals, they do not become infected.

Questions to ponder:

- Is it possible that prions could still be the culprits, despite the fact that the cloned protein mentioned in the scenario above wasn't infectious?
- How could a prion cause an infectious disease? Suggest a possible mechanism.
- Athough Dr. Stanley Prusiner was awarded the Nobel Prize in Medicine in 1997 for his work on prions, his discoveries are still regarded with skepticism by some researchers, and he is often referred to as a "maverick" scientist. Is this reluctance to accept new and controversial ideas a routine part of the process of science, or is this (Prusiner and prions) a special case?

Stage 3:
MAD COWS OF KENT

In 1996, the British government finally released a statement admitting that consumption of beef from mad cows was the most likely explanation for the recent appearance in England of a previously unknown form of human dementia. Christened "new variant Creutzfeldt-Jacob disease (vCJD)", it had by then contributed to the deaths of 14 (unusually young) individuals in the United Kingdom (*Nature* 385:197, 1997), whose brains upon autopsy had the characteristic spongiform degeneration and some more unusual neuropathologic features. These new cases more closely resembled the cow disease than they did the usual cases of Creutzfeldt-Jacob.

Meanwhile, back in the U.S.A., a 1996 popular press report (*Newsweek* 127[15]:58, 1996) stated that according to the USDA, 14 percent of cattle carcasses rendered each year were fed to other U.S. cattle. It has also been standard practice in this country to add rendered sheep tissue to cattle feed. This practice has since been banned by an FDA regulation that

went into effect in June 1997. A U.S. embargo on the importation of all cattle and cattle products originating in Great Britain has been in place since 1989, despite the fact that Britain has discontinued many of the practices thought to have been responsible for the mad cow disease outbreaks since 1988. No case of mad cow disease has yet been detected in the United States, but the annual incidence of scrapie in sheep remains about 30 to 50 cases per 8 million animals.

Questions to ponder:

- Now that these bans are in place in the United States, is it likely that mad cow disease will be a problem in this country? In your opinion, are there additional protective measures that need to be taken?
- Are you convinced that prions are the most likely transmissible agent of BSE and other transmissible spongiform encephalopathies? Why or why not?

Topics Introduced by the Problem:
MAD COWS OF KENT

Basic Background Topics

(Your textbook can be consulted for information on these.)
- Infectious processes in diseases of microbial and viral origin
- Immune system responses to viruses and bacteria
- Prions
- The structure and function of proteins
- How proteins are modified after they are made
- The normal structure and function of brain cells

More Advanced Topics

(You may need to consult additional resources for information on these.)
- Koch's postulates
- The epidemiology of BSE (and prion diseases of humans and sheep)
- Molecular biology of prions
- Husbandry of herd animals that humans consume
- FDA regulations concerning animal feed (animals for human consumption)
- Pathological changes associated with prion diseases

Additional Resources for Researching Problem Content:
MAD COWS OF KENT

....

Articles and books:

Caldwell, M. 1991. "Mad Cows and Wild Proteins." *Discover* April:69–74. (This account of the first major BSE outbreak was the principal source used to write the first version of this problem in 1993.)

Gibbs, C. J., Jr., editor. 1996. *Bovine Spongiform Encephalopathy: The BSE Dilemma*. New York: Springer.

Kolata, G. 1997. "Scientists Split over Prion Hypothesis." *New York Times* October 7:18.

Prusiner, S. B. 1996. "The Prion Diseases." *Scientific American* 272:48–55. (Also online at <http://www.sciam.com/0896issue/prion.html>.)

Rhodes, R. 1997. *Deadly Feasts*. New York: Simon and Schuster.

United States Congress, House Committee on Government Reform and Oversight. 1997. *Potential Transmission of Spongiform Encephalopathies to Humans: The Food and Drug Administration's (FDA) Ruminant to Ruminant Feed Ban and the Safety of Other Products: Hearing Before the Committee on Government Reform and Oversight, House of Representatives, One Hundred Fifth Congress, First Session, January 29, 1997*. Washington, D.C.: U.S. Government Printing Office.

Additional Resources for Researching Problem Content:
MAD COWS OF KENT

Electronic resources:

Baumbach, G. Department of Pathology, University of Iowa College of Medicine. "Infectious Diseases of the Central Nervous System Parenchymal Infections: Prions." *The Virtual Hospital.* <http://vh.radiology.uiowa.edu/Providers/TeachingFiles/CNSInfDisR2/Text/ PInf.CDE.html>

Brown, J. C. Department of Microbiology, University of Kansas. *What the Heck Is "Mad Cow" Disease?* <http://falcon.cc.ukans.edu/%7Ejbrown/madcow.html>

Brewer, S., J. Novakofski, and R. Wallace. University of Illinois at Urbana Champaign. *BSE—Bovine Spongiform Encephalopathy ("Mad Cow Disease").* <http://w3.aces.uiuc.edu/AnSci/BSE/>

FDA Center for Veterinary Medicine. *BSE Documents.* <http://www.cvm.fda.gov/fda/TOCs/bsetoc.html>

Thomasson, W. A. "Unraveling the Mystery of Protein Folding." *Breakthroughs in Science—FASEB.* <http://www.faseb.org/opar/protfold/protein.html>

University of Wisconsin. "Mad Cows: Behind the British Beef Scare." *The Why Files: Science Behind the News.* <http://whyfiles.news.wisc.edu/012mad_cow/>

World Health Organization. "Bovine Spongiform Encephalopathy (BSE) Fact Sheet (N113 March 1996)." *Emerging and Other Communicable Diseases (EMC).* <http://www.who.ch/programmes/emc/bsefacts.htm>

A MIX-UP AT THE FERTILITY CLINIC

Deborah E. Allen

"Did I just forget to change the pipette tip?" wonders Anke, a technician at a prestigious hospital-based fertility clinic in the Netherlands. It has been a slow day (only three in vitro fertilizations so far, the first involving a couple who came all the way from Aruba), and it has been difficult to keep her mind from wandering. "Oh well," she muses, "the tip always seems to clear out completely anyway. Changing it between samples is probably just an extra precaution."

Sixteen months later, in a small village in the same country, Wilma and Willem Stuart are beginning to feel like outcasts in their own community. Wilma and their two twins, Teun and Koen, have for the past several months been the subject of rather vicious gossip. For example, when Wilma goes out shopping with the twins in a stroller, people she encounters make comments such as, "What? Are those twins? How is that possible?" or "He is called Koen? Such a Dutch name for such an exotic child!" There was even a suggestion from one neighbor that Wilma had slept with another man. "Go ahead," this neighbor urged her. "Tell us your secret. Did you sleep with two men at the same time?" (M. Simons, "Uproar over Twins, and a Dutch Couple's Anguish," *New York Times*, June 28, 1995.)

Wilma and Willem also start to notice that Teun and Koen are different, although it takes some time after their birth for them to realize it. Koen's skin, slightly darker at birth, continues to grow darker as the months progress. They finally decide to consult their family physician when the twins are seven months old. He orders some tests and refers them to a psychotherapist for help coping with their own "bewilderment and pain" and to prepare them for what to tell one of their twins if the circumstances surrounding his birth are what they suspect: that a mix-up occurred on that slow day at the fertility clinic.

As they wait for the results of the tests ordered by their physician, Wilma and Willem consider whether they should consult a lawyer. They are hesitant because they aren't sure that the law will protect their interests, and they don't want to risk losing their child; they love him, and wouldn't dream of giving him up.

Questions to ponder:

- If you were this couple's family physician, what tests would you order? What samples would need to be collected from the concerned parties in order to run these tests?
- Who are the likely parents of the two children? What are the possibilities?

- Should Wilma and Willem have consulted a physician? Should they consult a lawyer?
- What advice would you have for a lawyer about how to represent their interests? Would the laws in the United States concerning parental custody rights protect a couple in a similar situation?
- Is it possible that a mix-up occurred? How could such a thing happen in a prestigious fertility clinic? What are the most likely possibilities? If one has occurred, what safeguards should the hospital take to prevent it from happening again?
- Is what the neighbor suggested possible in humans? That is, can there be two different fathers for twins born without technological intervention?

Topics Introduced by the Problem:
A MIX-UP AT THE FERTILITY CLINIC

....

Basic Background Topics

(Your textbook can be consulted for information on these.)
- Early embryonic development
- Fertility and fertilization (including in vitro)
- Menstrual cycle
- Formation of eggs and sperm

More Advanced Topics

(You may want to consult additional resources for information on these.)
- Paternity testing
- The process of ovulation
- Ethical issues arising from new reproductive technologies

Additional Resources for Researching Problem Content:
A MIX-UP AT THE FERTILITY CLINIC

Books and articles:

Kaplan, L. J., and R. Tong. 1996. *Controlling Our Reproductive Destiny*. Boston: MIT Press.

Strong, B., C. DeVault, and B. W. Sayad. 1996. *Core Concepts in Human Sexuality*. Mountain View, CA: Mayfield Publishing.

Weber, T., and J. Marquis. 1996. "In Quest for Miracles, Did Fertility Clinic Go Too Far?" *Los Angeles Times*, June 4. p. A1. (This is an account of another "mixup.")

Electronic resources:

American Society for Reproductive Medicine (ASRM) Homepage. <http://www.asrm.com/>

The InterNational Council on Infertility Information Dissemination (INCIID) Homepage. <http://www.inciid.org/>

University of California, San Francisco. *The Visible Embryo.* <http://visembryo.ucsf.edu/>

Stage 1:

SHOULD VICTORIA SKATE?

Deborah E. Allen

The International Olympic Committee (IOC) has convened for an impromptu meeting. Routine pre-competition blood screening has revealed that 16-year-old Victoria Boehmer, a member of the U.S. Women's Figure Skating Team, is actually a genetic male.

Although ill-timed injuries have prevented Victoria from competing at the top international level before now, she has won the U.S. national championship for the second year in a row. With her tall and lean body profile, balletic skating style, classic good looks, and charming smile, she has been a "media darling" and crowd pleaser since she first appeared on the ice. The media have reported that she has already lined up well over a million dollars worth of advertising contracts and other forms of corporate sponsorship. In short, Victoria is the odds-on favorite among the three top contenders for the gold (the other two are from the former Soviet Union).

An IOC member has already put in an unofficial and confidential phone call to Victoria's skating coach (also a close friend of her family), who is flabbergasted by the news. "I've been working with her since she was seven—she's no different from any other girl that I've coached. You've seen her—there's no way. There must be some mistake. This isn't something that Victoria or her parents could have hidden from me, even if they had wanted to. Besides, it would be so unlike them to try to pull a stunt like that."

Questions to ponder:

- Should the committee disqualify Victoria from the competition, or is there an alternate strategy?
- If the committee disqualifies her, how should they break the news to Victoria? To the press?
- Could Victoria be a genetic male without her or her parents being aware of it up until now? How would it be possible?

Stage 2:

SHOULD VICTORIA SKATE?

Results of this routine precompetition karyotyping prompted Victoria to seek medical advice. The physician she sees learns from her and her parents that she was born with an unambiguously female external phenotype, and has had a nearly uneventful childhood from a medical perspective, with the exception of an operation at age 8 to correct a unilateral inguinal hernia. Physical examination reveals normal mammary gland development, but an abnormally short vagina that ends blindly, consistent with her amenorrheic status. (Victoria has reported her amenorrhea previously to other physicians.) This initial exam reveals no other physical abnormalities. Reports of lab tests show levels of testosterone and of 5 alpha-reductase activity that are in the normal male range. With this information, the physician thinks he has a good explanation for her condition (he recently read an article that suggests it might be the result of point mutations in DNA). He doesn't want to tell Victoria and her parents what he suspects until he gets the results from some additional tests. In light of the lab information he has just received, he also needs to ask Victoria's parents some more questions about her hernia operation, and about the family's medical history.

Meanwhile, the manager of a professional ice skating company (that tours the country to put on glitzy shows) has contacted Victoria's coach.

"I'd like to offer Victoria a contract, and I'm touching base with you to see if you think she'd go for it. Given those rumors floating around that she's really a guy, we thought we could have her do a sort of "Victor/Victoria on Ice" routine. I know it's kind of "over the top," but we think there's a market out there for folks who want more than the usual family fare, and we'd make it worth her while."

The coach forces himself to make some barely polite replies, then hangs up abruptly. "I'm not sure I should even mention this to Victoria," he mutters to himself.

Questions to ponder:

- Why didn't the other physicians detect Victoria's condition before this?
- How could a person with a male genotype and normal testosterone levels be born looking like a female? What are the possible mechanisms?
- What additional tests or information might confirm one mechanism for Victoria's condition and at the same time eliminate the alternative ones?
- What do you envision is the long-term outlook for Victoria, from both a medical and a personal perspective? If you were a friend or family member from whom she was seeking advice, what would you say to her?

Topics Introduced by the Problem:
SHOULD VICTORIA SKATE?

• •

Basic Background Topics

(Your textbook can be consulted for information on these.)
• The basics of sex differentiation (from chromosomal sex to sexual maturation at puberty)
• Testosterone
• Male and female reproductive anatomy and function
• Karyotyping

More Advanced Topics

(You may want to consult additional resources for information on these.)
• Symptoms and mechanisms of abnormal sex differentiation
• 5 alpha-reductase
• Amenorrhea in young females

Additional Resources for Researching Problem Content:
SHOULD VICTORIA SKATE?

● ●

Articles and books:

Marieb, E. N. 1998. *Human Anatomy and Physiology.* Menlo Park, CA: Benjamin/Cummings.

Masters, W. H., V. E. Johnson, and R. C. Kolodny. 1995. *Human Sexuality.* Reading, MA: Addison-Wesley.

Strong, B., C. DeVault, and B. W. Sayad. 1996. *Core Concepts in Human Sexuality.* Mountain View, CA: Mayfield Publishing Company.

Electronic resources:

Intersex Society of North America Home Page. <http://www.isna.org/>

Health Answers—Healthy Living Online. <http://www.healthanswers.com>

Oregon Health Sciences University. "Sex Differentiation Disorders." *Cliniweb.* <http://www.ohsu.edu/cliniweb/C19/C19.391.775.html>

Your instructor may want to suggest some additional resources once you have had a chance to diagnose Victoria's condition.

Stage 1:

HUMAN IMMUNODEFICIENCY VIRUS (HIV) AND THE HEALTH-CARE PROFESSIONAL

Deborah E. Allen

After 16 years of serving as a congressman (U.S. House of Representatives), George Goodenough has now returned to his earlier profession as a lawyer specializing in medical malpractice claims. He has just ended a meeting with a client who claims that she was infected with HIV as the result of surgery (performed by an orthopedic surgeon) to correct an anterior cruciate ligament injury. George's meeting with his client has brought back memories of his former profession, in the course of which he served as a member of a House health subcommittee. On one memorable day, the committee was hearing testimony from several citizen panels concerning issues related to transmission of HIV in a health-care setting.

Kimberly Bergalis, a young woman who was thought to be the first victim of practitioner-to-patient HIV transmission, was first on the schedule to testify. She had been an ardent and vocal supporter of a bill sponsored by George's colleague, Rep. Dannemeyer (R–CA), that would make HIV testing mandatory for health-care workers who perform invasive procedures. Although the bill had not been scheduled for formal action by the subcommittee,

George and its other members had predicted that much of that day's public debate would center around its intent.

George can still remember the emotional impact of Ms. Bergalis's dramatic entry through a throng of reporters and cameramen. The room immediately grew as she began to make her brief statement to the subcommittee from her wheelchair.

"I'd like to say that AIDS is a terrible disease....I did nothing wrong, and I'm being made to suffer like this. My life has been taken from me. Please enact legislation so no other patient or health care provider will have to go through the hell that I have. Thank you."(The Baltimore Alternative 6:1,7; 1991)

Opposing arguments to the bill were presented by other speakers (including a nurse who contracted the virus from an emergency room patient, and an HIV-infected gay rights activist), but as George had suspected, news accounts of the hearing had focused almost exclusively on the drama of Ms. Bergalis's testimony.

George snaps out of his reverie, and realizes that he needs to have some more information before he can plan how to build his client's case. The Kimberly Bergalis story may well be relevant to his client's. He makes a mental note to ask his firm's researchers to "get on it" first thing in the morning.

Questions to ponder:

- What information do you think George should request from his firm's researchers?
- How commonly do physicians, dentists, and other health-care providers become infected with HIV? How does the incidence of infection compare to that of their patients? With their incidence of infection with hepatitis B?
- Are there currently any official restrictions on the professional activities of HIV-infected health-care workers? What restrictions, if any, do you think there should be?

Stage 2:

HUMAN IMMUNODEFICIENCY VIRUS (HIV) AND THE HEALTH-CARE PROFESSIONAL

Several days later George is pleased to find the requested information sitting in a folder on his desk, including a copy of the original Centers for Disease Control (CDC) case report that summarized the epidemiologic and laboratory findings of the investigation into the source of Kimberly Bergalis's HIV infection. As he scans through the first few sentences of the report, he is immediately struck by the contrast between its dry, dispassionate tone and that of the impassioned House subcommittee testimony dredged up out of his memory by his client's visit a few days ago.

The patient had two maxillary third molars extracted under local anesthesia in the dentist's office. The dentist had been diagnosed with AIDS 3 months before performing the procedure....She [the patient] did not recall, nor did review of the dental records reveal, any circumstances that would have exposed her to the dentist's blood....Four weeks after the dental procedure, the patient sought medical evaluation for a sore throat....The patient was diagnosed with pharyngitis and aphthous ulcers. Seventeen months after the procedure, she was diagnosed with oral candidiasis; 24 months after the procedure, she was diagnosed with Pneumocystis carini pneumonia and was seropositive for HIV antibody. (Morbidity and Mortality Weekly Report 39:489–493; 1990.)

This paragraph in particular prompts George to write a few questions in the margins for follow-up by the research team.

Questions to ponder:

- How did the researcher know that the CDC report he included in the file contained information about Ms. Bergalis's medical history? Does the CDC typically provide the names of the patients whose medical histories are summarized in its case reports?
- What modes of HIV transmission must have been eliminated from consideration before the possibility of Ms. Bergalis contracting HIV from an invasive dental procedure was considered? What questions should have been posed to the dentist and his employees once this possibility was considered?
- What other questions relevant to his client's case might this paragraph from the CDC case report have prompted in George?

Stage 3:

HUMAN IMMUNODEFICIENCY VIRUS (HIV) AND THE HEALTH-CARE PROFESSIONAL

As George reads further in the report full of tables and highly technical language, he suspects he will need to call in one of the researchers immediately to translate for him. First of all, he doesn't understand how samples of blood from Ms. Bergalis and the dentist were used to determine the relatedness of the HIV strains from both persons. In fact, he's not even sure why this information about relatedness was considered useful in determining the source of Kimberly Bergalis's HIV infection. His head reels as he encounters phrases such as, "peripheral blood mononuclear cells," "HIV sequences encoding the variable and constant regions of the major external glycoprotein, gp120," "pair-wise differences to sequences from corresponding DNA regions from 17 other distinct North American isolates."

"Why don't they just get to the point," George wonders in frustration as he tries to work his way through all the technical jargon. "Did her dentist give her AIDS, or didn't he? I need to know right away whether or not there's a precedent here for my client's claim."

George calls one of the researchers and ushers him into his office, hoping he can give him some simple answers. Instead, the researcher places a copy of a table he has compiled from several sources on George's desk. "I didn't get finished processing this soon enough to include it in the original report on your desk," he says eagerly. "It provides some information on 8 HIV-infected patients who have used the same dental practice as Kimberly Bergalis—I'm sure you'll find it useful."

George groans as he sits down to look at it, hoping the researcher will take pity on him and just tell him the right answer to his "bottom-line" question.

Questions to ponder:

- Why were DNA sequences coding for gp120 isolated and used to provide information about the relatedness of HIV strains in the patient and dentist?
- Now that you (and George) have considered some of the available evidence, do you think that George has the necessary precedent that will help him build his client's case?

Patient	Identified risk factor for HIV	CD4+/µL*	MHA-TP*	HBsAg*	DNA sequences closely related to sequences of dentist's virus	Amino acid signature pattern
A	No	<50	Negative	Negative	Yes	Yes
B	No	220	Negative	Not done	Yes	Yes
C	No	<50	Negative	Not done	Yes	Yes
D	Yes	Not done	Negative	Negative	No	No
E	No	560	Negative	Negative	Yes	Yes
F	Yes	250	Negative	Negative	No	No
G	No	400	Negative	Negative	Yes	Yes
H	Yes	640	Negative	Positive	No	No
Dentist	Yes	190**	Negative	Negative		Yes

MHA-TP = microagglutination assay for *Treponema pallidum* antibodies
HbsAg = hepatitis B surface antigen
*At start of epidemiologic investigation
**At time of AIDS diagnosis

Source: From *Annals of Internal Medicine* 116:799, 1992; and *Morbidity and Mortality Weekly Report* 40:378, 1991.

Topics Introduced by the Problem:
HUMAN IMMUNODEFICIENCY VIRUS (HIV) AND THE HEALTH-CARE PROFESSIONAL

Basic Background Topics

(Your textbook can be consulted for information on these.)
- General features of viruses, including structure and method of reproduction
- Reproductive cycles of animal viruses, with a focus on retroviruses such as HIV
- Modes of infection and transmission of viruses
- The human immune system

More Advanced Topics

(You may want to consult additional resources for information on these.)
- Human immune system responses to viral invasion
- Evolution of retroviruses
- Epidemiology of HIV/AIDS
- Physician/patient rights

Additional Resources for Researching Problem Content:
HUMAN IMMUNODEFICIENCY VIRUS (HIV) AND THE HEALTH-CARE PROFESSIONAL

Articles and books:

Fan, H., Conner, R., and L. P. Villarreal. 1996. *AIDS: Science and Society*. Boston: Jones and Barlett.

Gentile, B. 1991. "Doctors and AIDS." *Newsweek* 118 (July 1):48–57. (Contains the text of a letter written by Kimberly Bergalis to Florida health officials shortly before her death.)

Greene, W. C. 1993. "AIDS and the Immune System." *Scientific American* September:98–105.

Janeway, Charles A. Jr. 1993. "How the Immune System Recognizes Invaders." *Scientific American* September:72–79.

Varmus, H. 1987. "Reverse Transcription." *Scientific American* 257:56–64.

Electronic resources:

American Medical Association. *JAMA HIV/AIDS Information Center.* <http://www.ama-assn.org/special/hiv/hivhome.htm>

Centers for Disease Control and Prevention. *CDC National AIDS Clearinghouse.* <http://cdcnac.org/>

Stage 1:

PATTY'S PET PALACE

Deborah E. Allen

Shirley Sevin, a 45-year-old pet groomer, has worked for Patty's Pet Palace for the past few years. On a routine day at the Palace, Shirley shampoos, clips, and/or flea dips anywhere from eight to fifteen dogs.

For the past year, it has been increasingly more difficult for Shirley to get through the work day. She has been extremely tired, and has been having occasional headaches, blurred vision, and bouts of nausea and dizzy spells. And, to top it off, she has begun to sweat profusely (especially at night) and often has trouble getting to sleep.

Shirley has kept her good friends apprised of her "symptoms of the week," and they have been giving her all sorts of advice about what could be wrong. "Sounds like allergies," one friend opined. "I hate to bring it up Shirl, but you're just getting older, and believe me, it's all downhill from here," another friend offered (not so?) helpfully. Advised yet another one, "I have just three little words to tell you—estrogen replacement therapy. It turned my life around." Another woman (who was never one of Shirley's favorites) was overheard to say, "Let's face it, girl, this is all in your head. You need to go see a shrink and unload all that excess bag-gage about your ex—the way that man treated you would make anyone sick."

One night at a local eating establishment, Shirley is telling her dinner companions of how her symptoms have escalated. She has begun to have periodic chest pain and shortness of breath, and during the past week she blacked out several times. In the middle of her tale, she has another brief blackout, and her friends are disturbed to see that her pupils (particularly the right eye) are narrowed to pinpoints. "Don't mess around with this any longer," they plead. "We had no idea it was this serious. Get yourself to a doctor."

Questions to ponder:

- Should Shirley wait until morning to seek medical assistance, or head right to an emergency room?
- What questions should a physician ask Shirley?
- Based on her symptoms, what body systems may be involved in her problems?
- Given the knowledge they had of Shirley's earliest symptoms, were her friends' original suggestions about what's wrong with Shirley reasonable ones?

Stage 2:
PATTY'S PET PALACE

....

Shirley goes in to see her family practitioner the next day, and because of her blackouts, she is referred to a neurologist. The neurologist orders a brain scan and electroencephalogram as diagnostic tests. Unfortunately, the tests fail to reveal what's causing her symptoms.

The day after her last visit to the neurologist, in a fortuitous turn of events, Shirley receives a call from a representative of her state's Health Assessment and Information Resource Network (HAIRNet). "We're conducting a survey of pet groomers in major cities in the state," the representative said. "Have you been experiencing any of the following symptoms?" When these symptoms (recently reported to HAIRNet by another pet groomer) read like a road map of Shirley's, Shirley immediately asks for advice about what to do, because her symptoms have not gone away.

Shirley made an appointment the next day with a physician who specialized in occupational medicine. Subsequent tests revealed red blood cell cholinesterase levels close to the limit of the normal range. Shirley was treated with atropine, and advised to "lay off" the flea dipping for a while. After several months of avoiding exposure to the flea dip,

Shirley's red blood cell cholinesterase levels were in the mid-normal range and most of her symptoms had disappeared, confirming the diagnosis of the occupational medicine specialist. The specialist has linked the symptoms to Shirley's treatment of her canine clients with a flea dip that she mixes up from a concentrate containing 11.6 percent phosmet.

Shirley faces being fired from her job, however, if she is not able to return to performing one of the staple services in the Pet Palace's full line for the devoted (but busy) dog owner— flea control.

Questions to ponder:
- Why was atropine the drug chosen for treating Shirley's symptoms?
- How does a substance designed to kill fleas cause harm to humans? Are we really that alike?
- How do you think the flea dip entered Shirley's bloodstream? What are the most likely routes of entry?
- Are there other possible sources of exposure to toxins (i.e., other than occupational exposure to flea dip) that might have been seriously considered in Shirley's case?
- What advice would you have for Shirley if she decides to reenlist for the flea patrol?

Topics Introduced by the Problem:
PATTY'S PET PALACE

....

Basic Background Topics

(Your textbook can be consulted for information on these.)
* Major divisions of the nervous system
* Anatomical and functional elements of a simple nervous pathway
* Information transfer in the nervous system
* Neurotransmitters
* The insect nervous system

More Advanced Topics

(You may want to consult additional resources for information on these.)
* Acetycholinesterase
* Atropine and its action
* Organophosphate pesticides
* The electroencephalogram and "brain scan" as diagnostic tests
* Estrogen replacement therapy
* Stress-induced illness
* Allergies

Additional Resources for Researching Problem Content:
PATTY'S PET PALACE

....

Articles and books:

Bardin, P. G., S. F. van Eeden, J. A. Moolman, A. P. Fodem, and J. R. Joubert. 1994. "Organophosphate and Carbamate Poisoning." *Archives of Internal Medicine* 154:1433–1441.

Burtis, C. A., E. R. Ashwood, and N. W. Tietz. 1998. *Tietz Textbook of Clinical Chemistry.* Philadelphia: W. B. Saunders.

Coye, M. J., J. A. Lowe, and K. T. Maddy. 1986. "Biological Monitoring of Agricultural Workers Exposed to Pesticides: I. Cholinesterase Activity Determinations." *Journal of Occupational Medicine* 28:619–627.

Guidotti, A. 1995. *Manual of Acute Pesticide Toxicity.* Boca Raton, FL: CRC Press.

Kamrin, M. A. 1997. *Pesticide Profiles.* Boca Raton, FL: CRC Press.

Electronic resources:

Extension Toxicology Network. "Pesticide Information Profiles." *EXTOXNET.* <http://ace.ace.orst.edu/info/extoxnet/pips/ghindex.html>

Office of Pesticide Programs, Environmental Protection Agency. *Pesticide Product Information System.* <http://www.epa.gov/opppmsd1/PPISdata/index.html>

USDA Agricultural Research Service. *Agriculture Network Information Center Home Page.* <http://waffle.nal.usda.gov>

University of Florida, Institute of Food and Agricultural Science. *FAIRS— Florida Agricultural Information Retrieval System Home Page.* <http://hammock.ifas.ufl.edu>

Stage 1:

AN ANDEAN ADVENTURE

Deborah E. Allen

At last! Noah Clooney thought he'd never survive, but here he is, fresh out of medical school, on his way to southern Peru to join a team of professionals that has been providing free medical care to remote Andean villages. As his plane lands in Cusco, he is greeted by the physician with whom he has been in contact since he first decided to "take a breather" before starting his internship.

She greets Noah rather hastily, hurries him into a waiting vehicle, and as she's negotiating the early morning traffic, she offers an apologetic explanation."I'm afraid you're about to find out what we're all about firsthand. Just before I left for the airport an *indigena* (a member of the Quechua population) came to the clinic with some disturbing news. There's been an outbreak of a serious illness in his *allyu*, and immediate medical attention is needed. Members of our team have already started out, but I'm sure they're going to need my help as well. Would you like to go along and observe what we do?"

Noah nods his assent, slightly offended that she doesn't seem to think he'll be of any help to the sick villagers. "Maybe she's assuming that I'll need more time to adjust, and doesn't realize that my experience from skiing trips in the Rockies makes me sure I won't have any problems," Noah says to calm himself down.

After a long trip through some picturesque scenery (he particularly enjoyed seeing the (vicuñas), Noah and the more senior physician finally arrive in a small town in the district of Nuñoa, where they load their supplies in packs that will be carried by a llama. A guide will escort them the remaining two miles to the sick villagers.

The following morning, Noah has come around to his fellow physician's point of view. As he drags himself out of bed, he's acutely aware that he has a splitting headache, is nauseous, and is completely exhausted from having spent an almost sleepless night. He has a vague and disquieting recollection of having staggered off the path, nearly falling off a cliff and down the side of the mountain. ("Did the guide really grab a handful of my shirt in the nick of time?" he muses, and wonders if he'll get a chance to thank him.) He recalls the visions of curiously barrel-chested villagers that danced through his head as he lay awake last night— were they real, or some strange hallucination? "I never felt this way in the Rockies," he reflects with a tinge of chagrin. How in the world do the villagers (and the vicuñas and llamas, for that matter) manage to live here permanently?

Questions to Ponder:

- What aspect(s) of the high-altitude environment pose a problem for humans and other organisms?
- Do the villagers have genetic differences (from lowlanders) that allow them to live permanently at these high altitudes?

- Why would high-altitude exposure result in such a collection of seemingly unrelated symptoms?
- What treatment will Noah's new colleague (the other physician) recommend for him?

Stage 2:
AN ANDEAN ADVENTURE

For Noah, this "Andean Adventure" turned out to be a life-transforming experience. Upon returning to the States, he made a quick about-face, deciding to subspecialize in high-altitude medicine. Now, 15 years later, he has been asked to serve as a consultant to a team of professional bicyclists that is in training for an upcoming international race to be held at 9,000 feet near Bogota, Colombia. His excellent reputation, along with an avid interest in the sport (he has successfully competed in several local amateur races), made Noah the top choice for the job. He's now helping the team to formulate an optimal training regimen.

The team receives financial support from several private individuals and corporate sponsors, and will be awarded a substantial cash prize if it wins the race. The sponsors will receive considerable free advertising during the prestigious and well-known event, and even more afterward if the team is successful. Late one afternoon, Noah receives a call from a representative from one of these companies, an "up-and-coming" biotech firm that is the team's biggest sponsor.

"I know how we can get hold of some EPO [NB: *erythropoietin*]," the representative informs him as soon as the "hellos" are out of the way. "Just tell me how soon and how often the team needs to get it, and I'll have my contact take care of it—I wouldn't want you to take the heat if someone finds out. Bad for PR and all that. There's no way the [race] rules committee will be able to screen for it, so we'll be OK from that perspective. It should give our team a real edge over the competition."

Noah, somewhat taken aback by the request, thinks carefully before he crafts his response.

Questions to ponder:

- Should the team consider moving to a town close to the Sierras or Rockies to train for the race, or is there a better strategy?
- What advice do you have for Noah about how to handle this situation?

Topics Introduced by the Problem:
AN ANDEAN ADVENTURE

· ·

Basic Background Topics
(Your textbook can be consulted for information on these.)
• Homeostasis and the body systems that help maintain it
• Adaptation to differing environmental conditions
• Respiratory gas exchange
• Oxygen and carbon dioxide transport in the blood
• Body responses and adjustments to exercise

More Advanced Topics
(You may need to consult additional resources for information on these.)
• The oxygen dissociation curve for hemoglobin and the factors that influence the extent of oxygen binding to hemoglobin
• How the body controls breathing, particularly chemoreceptor reflexes
• Mechanisms to enhance the delivery of oxygen to tissues
• Adjustments (acclimatization) to high-altitude exposure
• Erythropoietin
• Types of muscle fibers
• Conditioning for competitive sports

Additional Resources for Researching Problem Content:
AN ANDEAN ADVENTURE

··

Articles and books:

Houston, C. S. 1992. "Mountain Sickness." *Scientific American* October:58–66.

Marieb, E. N. 1998. *Human Anatomy and Physiology*. Menlo Park, CA: Benjamin/Cummings.

Perutz, M. F. 1978. "Hemoglobin Structure and Respiratory Transport." *Scientific American* December:92–125.

Rhodes, R., and R. Pflanzer. 1996. *Human Physiology*. Philadelphia, PA: Saunders.

Electronic resources:

Curtis, R. Outdoor Action Program, Princeton University. "High Altitude: Acclimatization and Illnesses." *Durango Downtown Home Page.* <http://www.creativelinks.com/recreat/altitude.htm>

Coghlan, A. "Human Cells Make 'Perfect' Proteins." *New Scientist—Planet Science.* <http://warming.newscientist.com/ns/970503/transkar.html>

Magnay, J. "Drawing Blood in the War against Drug Cheats." *The Sidney Morning Herald.* <http://www.smh.com.au/daily/content/970822/pageone9.html>

University of Mississippi Medical Center, Department of Physiology and Biophysics. "QCP2 Lab—Oxygen Homeostasis." *Quantitative Circulatory Physiology Lab.* <http://phys-main.umsmed.edu/studlabs/QCP2LABS/O2HOMEOS/O2HOMEOS.HTM>

Stage 1:
ZÖE TAKES A DIVE
Deborah E. Allen

....

It's austral summer outside, but Steve's not feeling quite as lucky as he did several months ago about having the opportunity to spend his extended holiday break assisting a research team on a National Science Foundation–sponsored expedition to Ross Island, Antarctica. The group he's working with, headed by his faculty advisor from Richter State U., has some new telemetry devices he's been attaching to Weddell seals (*Leptonychotes weddelli*). These devices have enabled him and the team to monitor heart rate, body temperature, blood pressure, cardiac output (blood flow in the ascending aorta), and organ blood flows during the seals' descents and ascents from voluntary dives. They also send signals from a depth recorder attached to the animals.

Right now, Steve is cold, hungry and miserable, having spent several hours in a small hut over a manmade hole in the sea ice about 15 kilometers offshore. He and Samantha (another student on the expedition) are waiting for their latest subject (who they've christened "Zöe") to ascend from a dive. She has been down about 50 minutes already, and when she ascends Steve and Samantha will still have about 3 hours of analytical work to do on the post-dive blood samples they'll have to collect from her. Just as Steve and Sam are about to head back to the base camp, thinking "Zöe" must be dead, he notices her vibrissae beginning to emerge from the water.

Questions to ponder:

- What physiological and physical problems must the seal have faced in the course of this lengthy dive?
- Why didn't the seal get the "bends," despite the relatively rapid ascent that Steve's telemetry equipment has recorded? Wouldn't a human diver have a problem with an ascent this fast?

Stage 2:
ZÖE TAKES A DIVE

The next day, his misery forgotten, Steve has some results he is eager to show his advisor. He stayed up late using the facility's computer to graph some of the data he and Samantha have been collecting, and for some of the biochemical data, he has noticed some interesting differences between yesterday's dive and a previous one (also by Zöe) of shorter duration.

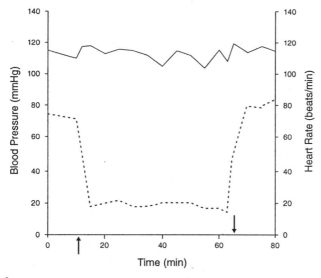

A   — Blood Pressure   - - Heart Rate

Figure 1A. Heart rate and mean arterial blood pressure before, during, and after a free dive to 512 meters depth in a Weddell seal. Arrows mark the start and end of a 52 minute dive in a pregnant female seal weighing 385 kg. The seal's body temperature (not shown) fell from 37.9 to 34.7 °C during the dive.

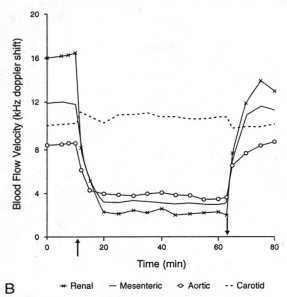

B   ✶ Renal   — Mesenteric   ⊶ Aortic   -- Carotid

Figure 1B. Blood flow velocities in several major arteries before, during, and after the same free dive to 512 meters depth in a Weddell seal. Arrows mark the start and end of the dive.

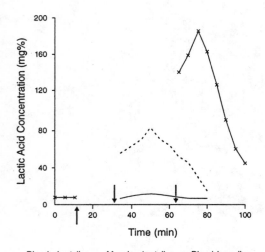

— Blood, short dive   - - Muscle, short dive   ✶ Blood, long dive

Figure 2. Lactic acid concentration of blood samples taken at intervals before and during recovery from a 19-minute duration (short dive) and a 52-minute duration (long dive) free dive in an adult female Weddell seal. Both dives start at the point on the time (X) axis indicated by the upward arrow; the first dive ends at the first downward arrow, and the second dive ends at the second downward arrow. Muscle lactate concentration was also measured during recovery from the shorter dive (dotted line).

Questions to ponder:

- Why might these cardiovascular changes (in Figure 1) be useful to a diving seal? Would a diving human experience these same changes? (After all, we don't have to dive to get the fish we eat.)
- Steve and the team weren't set up to measure lactic acid during the dive, but he has some good ideas (based on the pre- and postdive values) about how it must have changed.

  Complete the graph (Figure 2) by drawing in the curve for what you think happened to blood and muscle lactic acid concentration during both the shorter and longer dives.
- Although ventilatory rate was not measured, how would you anticipate that it would change (from the seal's normal breathing pattern before the dive) during recovery from the long dive? What mechanisms would account for the change?

Stage 3:
ZÖE TAKES A DIVE

Steve is back in his dorm at Richter State U., contemplating the rough draft of a paper his advisor has suggested he write as part of an independent study project. (Sam is going to write up the cardiovascular data.) Steve is trying to compare some of the biochemical information the Antarctic research team collected for Weddell seals with information he has been able to piece together from the scientific literature on other mammals, including some other pinnipeds and some cetaceans. He has gotten as far as drawing a graph of the data for the most likely predive conditions (thinking this would be a helpful way to organize his thoughts), but not for how the data determined during diving conditions might look. He is hoping that the comparative data will support his ideas about how diving mammals have adapted to the periodic need to survive oxygen deprivation. An interesting question pops into mind, however, as he thinks of that cold and miserable day that he and Samantha waited for Zöe the seal to ascend from her 52-minute dive.

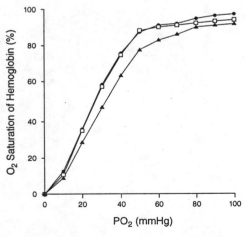

→ Humans ▫ Seals ▲ Pilot whales

Figure 3. Oxyhemoglobin dissociation curves compared for humans (solid circles), pilot whales (solid triangles), and seals (unfilled squares). The curve for each species is a composite of averaged data from several studies. In all cases, the curves were determined at a temperature of 37 °C and a pH of 7.4.

Table 1. Comparison of properties of
blood from adult Weddell seals and
humans

Blood Property	Seal	Human
Hematocrit (%)	55	45
Hemoglobin, tetramer (mmol/L)	3.3	2.3
Red blood cell DPG (mmol/L)	6.5	4.1

Questions to ponder:

- Why do you think Steve thought it was important to report (in Figure 3) the conditions under which the oxyhemoglobin dissociation curves he plotted were determined? Why should this matter?
- Help Steve to finish the graphs—that is, what might the curve for the seal look like during conditions more nearly similar to those during a dive?
- Using information provided in Table 1, how would the amount of hemoglobin found in each red blood cell (the mean corpuscular hemoglobin concentration) compare in the two species?
- What important difference between seal and human blood properties is missing from Steve's Table 1?
- Steve's thought was about Zöe's (near-term) fetus—do you think it survived her deep and prolonged dive? Why or why not?

Topics Introduced by the Problem:
ZÖE TAKES A DIVE

••

Basic Background Topics

(Your textbook can be consulted for information on these.)
- Basic respiratory processes of ventilation, gas exchange, and transport in vertebrates with lungs
- Basic anatomy of the mammalian circulatory and respiratory systems
- Control of heart function
- Blood vessels and blood flow
- Reflexes/reflex arcs
- Hemoglobin and myoglobin
- Anaerobic respiration (lactic acid fermentation)
- Components and function of blood

More Advanced Topics

(You may want to consult additional resources for information on these.)
- Factors that alter the binding of oxygen to hemoglobin
- The diving and chemoreceptor reflexes
- Special adaptations for diving in the Weddell seal
- Decompression sickness and other problems that divers face

Additional Resources for Researching Problem Content:
ZÖE TAKES A DIVE

Articles and books:

Bookspan, J. 1996. *Diving Physiology in Plain English.* Kensington, MD: Undersea and Hyperbaric Medical Society.

Boxer, S. 1985. "Why Seals Don't Get the Bends." *Discover* 6:11.

Broad, W. J. 1997. "Deep Underwater, the Breath of Life; Researchers Observe Diving Mammals for Clues to Improving Human Health." *The New York Times* 147 (November 11):B9, F1.

Clenney, T. L., and L. F. Lassen. 1996. "Recreational Scuba Diving Injuries." *American Family Physician* 53:1761–1767.

Kooyman, G. L., and P. J. Ponganis. 1997. "The Challenges of Diving to Depth." *American Scientist* 85 (November–December):530–539.

Melamed, Y., A. Shupak, and H. Bitterman. 1992. "Medical Problems Associated with Underwater Diving." *The New England Journal of Medicine* (January 2):30–35.

Zapol, W. M. 1987. "Diving Adaptations of the Weddell Seal." *Scientific American* 256:100–105.

Electronic resources:

U. S. Department of Commerce/National Oceanic and Atmospheric Administration. *NOAA Diving Manual. Table of Contents.* <http://uwsports.ycg.com/reference_library/noaa/>

U.S. Department of Commerce/National Oceanic and Atmospheric Administration. *National Marine Mammal Laboratory Home Page.* <http://nmml01.afsc.noaa.gov/>

Stage 1:
WATER, WATER EVERYWHERE...
Deborah E. Allen

. . . .

Maria, Alicia, Carlos, and Ben were born to sail. For the past ten years, they have spent all of their spare time and money cruising the lakes, rivers, bays, and other waterways within two week's distance and back from their home port. You name it, they've sailed it—with one exception. A cruise around the world has always been just beyond their grasp—until now.

After months of preparation, a few corporate sponsors (who signed up for advertising space on the Web site a sailing magazine has devoted to their exploits), and extended leaves of absence from their jobs, they have set out in a 45-foot ketch. This tale begins as the four compadres are somewhere in the middle of the Atlantic (check out the map on their Web site if you are particular about the details). It's the middle of the night, and they have just heard a small craft warning on their radio. A storm front is moving in.

The storm picks up, and it is a particularly vicious one, with waves towering over their sailboat. You guessed it—the crew is forced to abandon ship. They manage to get off the floundering sailboat and into the lifeboat just in the nick of time. In fact, Ben's departure is hastened by the boom, which makes one last sweep about to knock him unceremoniously into the smaller craft. Their best laid plans have also gone awry— the "abandon boat bags" that they have so carefully prepared get abandoned with the boat in the hasty scramble for the lifeboat.

A few days later, having brought no food and only one small container of water along with them, the castaways are thankful that they have managed to survive the storm, but they are in pretty bad shape. They managed to transmit a last radio message before abandoning ship, and hope that the Coast Guard will find them soon.

Carlos, with not much else to do, considers whether he wants to risk drinking some of the seawater. "I've heard you're not supposed to drink it, but I wonder if that's just an old wive's tale. "After all," he reasons, "those birds we saw this morning seemed to be scooping up the water, and I know that baleen whales (aren't they mammals, too?) must drink some seawater with the way they have to eat." Carlos and Alicia have insisted that Maria, who is three months pregnant, have the lion's share of what little water they had with them, and it is already long gone.

Question to ponder:
- Should Carlos and the others drink the seawater? Why or why not?
- Are there other options (other than drinking seawater) that might be available to them as a way to cope with their limited fluid intake?

Stage 2:
WATER, WATER EVERYWHERE...

Alicia interrupts Carlos's reverie by once again worrying out loud about Ben (her significant other). "Why is he still unconsciousness?" she wails. "And what's that pool of fluid that keeps forming under him? Is he peeing on himself? We better get help soon, or I'm afraid he's going to die."

Carlos holds back a retort that Ben might not be the only one to die.

But this story has a happy ending. The castaways are rescued by the Coast Guard and airlifted to the nearest hospital emergency room. One of the on-call physicians (Dr. Rollins, who is just out of medical school) measures their vital signs, orders a series of tests, and starts them all on intravenous fluid replacement.

Measurement	Range in Normally Hydrated Person	Alicia	Maria	Carlos	Ben
Serum Na+ (mEq/L)	136–144	155	152	155	164
Hematocrit (%)	41–50, males 3644, females	49	46	55	59
Serum Osmolality (mOsm/kg)	280–295	315	312	316	347
Serum Creatinine (mg/dl)	0.8–1.4	6.5	7.5	7.0	12.5
BUN (mg/dl)	7–20	55	50	62	118
Urine Osmolality (mOsm/kg)	500–800	1225	1250	1340	290
Body Weight (kg)	—	54	56	72	66

However, Dr. Rollins is puzzled by the test results for Ben. She shows them to the resident in charge (Dr. Gellai), hoping he has seen something like this before.

Dr. Gellai scans Ben's test results, and thinks for a few minutes. "This is odd...if I hadn't gone to the medical school that I did, I might not have even thought about this as a possibility. You should get the patient upstairs right away for a cranial MRI. It's a long-shot, but I think I'm right...also, have the pharmacy send some desmopressin acetate over, just in case."

Questions to ponder:

- What aspect of Ben's test results makes Dr. Gellai think that Ben might require special treatment?
- What type of fluid did Dr. Rollins most likely recommend for intravenous replacement in the case of these dehydrated castaways? If Dr. Gellai is right about his hunch, does this fluid replacement procedure have to be altered in any way for Ben?
- Physicians can quickly estimate the degree of water deficit and thus provide a guide for the fluid replacement needed for dehydrated

male patients using the following formula (that actually calculates the amount of water that must be added to return the plasma sodium concentration to a normal value of 140 mEq/L):

Body water deficit (liters) = 0.6 × current body weight in kilograms × [(Current serum Na+/140) − 1].

The formula is the same for *females*, except that the current body weight is multiplied by 0.5. The first part of the formula, 0.6 or 0.5 times the current body weight in kilograms, is actually an estimate of the current *total body water* in liters.

Given that physicians classify mild dehydration as a loss of 5 percent of total body water, moderate dehydration as a 10 percent loss, and severe dehydration as a 15 percent loss, how would you categorize the degree of dehydration in each of the four castaways? Is this an accurate estimate for Maria? Why or why not?

Basic Background Topics

Topics Introduced by the Problem:
WATER, WATER EVERYWHERE...

(Your textbook can be consulted for information on these.)
- How the kidneys work
- Homeostatic control mechanisms—a general scheme for how they work
- How the body gains and loses salt and water
- Hormones that help regulate the body's salt and water content
- The structure and function relationships of the endocrine glands involved in control of salt and water balance

More Advanced Topics

(You may want to consult additional resources for information on these.)
- The effects of dehydration on the body's systems and on the body's fluid compartments (humans and other mammals)
- Tests and measurements for assessing the body's fluid and electrolyte status, and the status of the cardiovascular system and kidneys

Books:

Additional Resources for Researching Problem Content:
WATER, WATER EVERYWHERE...

Buskirk, E. R., and S. M. Puhl, editors. 1996. *Body Fluid Balance: Exercise and Sport*. Boca Raton, FL: CRC Press.

Metheny, N. M., editor. 1996. *Fluid and Electrolyte Balance*. Philadelphia: Lippincott-Raven.

Rhodes, R., and R. Pflanzer. 1996. *Human Physiology*. Philadelphia: Saunders.

Schmidt-Nielsen, K. 1997. *Animal Physiology*. New York: Cambridge University Press.

Vander, A. L. 1994. *Renal Physiology*. New York: McGraw-Hill.

Weldy, N. J. 1995. *Body Fluids and Electrolytes*. Saint Louis: Mosby–Year Book.

Electronic Resources:

Coalition of Family Physicians of Ontario. *RxMed: The Website for Family Physicians.* <http://www.rxmed.com>

Orbis Broadcast Group. *Health Answers: McCalls/Family Circle—Healthy Living OnLine.* <http://www.healthanswers.com>

Your instructor may suggest some articles and additional Web sites about Ben's condition once you have diagnosed it.

Stage 1:

BEAT THE CLOCK?

Deborah E. Allen

Jeff Pyneil and his girlfriend Delia Zeitgeber are having lunch with her friend Helen Sungorus at a coffee bar in a big midwestern city. Helen has just volunteered to be a subject in a study some scientists at the local medical center are conducting on circadian clocks, and she's telling Jeff and his girlfriend all about it.

"I volunteered because I think I have this thing called seasonal affective disorder. I get all depressed and sleepy in the winter—sometimes I can barely pull myself out of bed in the morning. Someone told me the cure is to sit under bright lights with your eyes wide open for a few hours each morning. The guys doing the study are looking at whether you can get the same effect by just shining a light on the back of someone's knee while they're asleep! The downside is that they're going to make me sleep in a special room so they can measure my brain waves (electroencephalogram) and body temperature, and take blood to measure my melatonin levels. It's only for four days, though, so I guess I'll survive."

"Melatonin—isn't that the stuff that you say your store keeps running out of ever since those news stories on how it prevents aging came out?" Jeff asks Delia, who works in a natural foods store.

"Yeah, that's the stuff. I wonder why they're going to measure it in your blood. I thought it's mostly just a pill that people take to try to stop getting old—you know, turn back the clock. I'm not so sure it does anything other than make them think they feel younger, but my boss is selling a lot of it, so I keep quiet. Some of our customers take it for jet lag, though, and really swear by it."

"Why don't we run a little test of our own?" suggests Jeff. "I could take it on my trip to Hong Kong. I still have about four days before I leave. Could you bring me some from the store before I go? On my last trip to Europe, it took me almost five days to adjust, and it was really tough trying to stay awake at all the meetings I had to attend. I'd try just about anything to prevent that from happening on this trip."

Delia brings Jeff a bottle of the 2 milligram dose of a mix of synthetic and natural melatonin the following day. Jeff starts taking it each morning before his flight (with his coffee). On the flight west, he "pops" a few more pills with the beers he has to ease his boredom and anxiety about flying. While in Hong Kong, he continues to take it faithfully before he leaves for his business meetings each morning, but gives it up on the fourth day.

"It didn't work," he tells Delia when she meets him at the airport on his return. "I was more tired than ever," he complains. "At one point during a business dinner on the second day, my boss had to nudge me to keep me from falling face down in my soup. You must have some really whacky customers at that store of yours. I hope that none of them are airline pilots depending on this stuff to stay awake during flights."

Questions to ponder:

- How could popliteal (back of the knee) illumination conceivably alle-viate symptoms of winter depression? What body systems or pathways might be involved?

- Why were the scientists measuring brain waves, body temperature, and melatonin as part of the study? How might they intend to use this information in evaluating whether their hypothesis was supported or not?

- Did Delia give Jeff a "bum steer" about melatonin and jet lag, or should Jeff repeat his test on the next business trip he takes?

Topics Introduced by the Problem:
BEAT THE CLOCK?

Basic Background Topics in Biology

(Your textbook can be consulted for information on these.)
• Biological (circadian) rhythms
• Biological clocks
• Melatonin
• Pineal gland

More Advanced Topics

(You may want to consult additional resources for information on these.)
• Jet lag
• More in-depth look at all of the topics above

Additional Resources for Researching Problem Content:
BEAT THE CLOCK?

Books and articles:

Binkley, S. 1998. *Biological Clocks.* Newark, NJ: Gordon & Breach Publishing Group.

Brezinski, A. 1997. "Melatonin in Humans." *The New England Journal of Medicine* 336:186–195.

Cowley, G. 1995. "Melatonin Mania." *Newsweek* 126:60–63.

Keister, E. Jr. 1997. "'Traveling Light' Has New Meaning." *Smithsonian* 28:100–116.

Moore-Ede, M. C., F. M. Sulzman, and C. A. Fuller. 1984. *The Clocks That Time Us.* Cambridge, MA: Harvard University Press.

Oren, D. A., and M. Terman. 1998. "Tweaking the Human Circadian Clock with Light." *Science* 279:333–335.

Reiter, R. J., and J. Robinson. 1996. *Melatonin: Your Body's Natural Wonder Drug.* New York: Bantam Books.

Electronic resources:

Columbia University Health Services, Healthwise. *Go Ask Alice.* <http://www.columbia.edu/cu/healthwise/about.html>

Hsiung, R. *Dr. Bob's Psychopharmacology Tips.* <http://uhs.bsd.uchicago.edu/dr-bob/tips/>

Nature's Life. "Melatonin." *Nature's Life Web Site.* <http://www.natlife.com/Product/melatonin.html>

Stage 1:
MARTHA WANTS A BLUE RIBBON AT THE FLOWER SHOW (OR, KEEPING MUMS)
Deborah E. Allen

Martha Graham just inherited an old Pennsylvania farmhouse with 50 acres of surrounding fields from her grandmother. Her grandmother used to operate a nursery on the premises, and Martha hopes to continue the business once she is able to get a bank loan to finance her starting costs and some much-needed repairs to the greenhouses.

Martha hopes to keep the nursery afloat by cultivating a line of flowering plants for sale in the houseplant section of the local supermarkets. She particularly is interested in cultivating showy, fall-flowering plants (maybe chrysanthemums) that she will time to flower to hit the market in the midst of the winter doldrums. She hopes to introduce the line with a splash by exhibiting (and hopefully winning some ribbons) at this year's Philadelphia Flower Show.

She has learned a lot from her grandmother, but needs to do some additional research on how to time the blooming of the plants (with minimal use of energy), what growth medium and fertilizers to use,

whether judiciously timed use of plant hormones would be more trouble than it's worth, and whether she will need to use pesticides and herbicides to get an optimal yield (also, are they safe for her to use?). Unfortunately, the bank expects Martha to have a handle on all of this information because she has to submit a business prospectus as part of the application procedure for her commercial loan.

You were drawn into the world of Martha's mums one evening when you met her at a party and she learned that you are an extension specialist in the agriculture college of a local university.

Questions to ponder:

- What advice would you have for Martha about how to get started?
- Should she use plant hormones to promote growth and flowering, or are their better ways to do this?
- What conditions should Martha maintain in her greenhouse if she wants to have spectacular mums to display at the flower show?

Stage 2:

MARTHA WANTS A BLUE RIBBON AT THE FLOWER SHOW (OR, KEEPING MUMS)

You offer Martha some sound advice about how to plan the start-up phase of her nursery operation. Once she has the business up and running, she calls you for help with problems that arise from time to time, as she has this afternoon.

Once you have given her some advice about her mums' "disease-of-the-month," she asks you about some new technological advances in plant genetics and biotechnology that she read about in a plant growers' magazine. One advance is the discovery of the so-called "S" gene; manipulation of its expression might lead to a method for controlling self-fertilization, always a concern for commercial growers who use hybrid seeds.

You have heard of this work, and point out to Martha that plant biotechnologists are also beginning to make use of antisense RNA in ways that might one day enable her to grow "whiter than white" mums.

Martha is impressed with both the techniques and your understanding of how they work, but wonders if these new "high tech" methods will ever be practical for a small nursery operation such as hers.

Questions to ponder:

- Why would manipulation of the "S" gene be useful to large-scale commercial plant growers?
- How might Martha use antisense RNA to make "whiter than white" mums? Will someone like Martha be able to use this technology in the near future?

Topics Introduced by the Problem:

MARTHA WANTS A BLUE RIBBON AT THE FLOWER SHOW (OR, KEEPING MUMS)

Basic Background Topics

(Your textbook can be consulted for information on these.)
- The responses of plants to the environment
- The life cycle of angiosperms
- Flowers and flowering
- Plant hormones
- Developmental stages of plants
- Self- versus cross-fertilization

More Advanced Topics

(You may want to consult additional resources for information on these.)
- Nutritional requirements of plants
- Antisense RNA
- Plant "S" genes

Additional Resources for Researching Problem Content:
MARTHA WANTS A BLUE RIBBON AT THE FLOWER SHOW (OR, KEEPING MUMS)

Books:

Chrispeels, M. J. 1994. *Plants, Genes and Agriculture*. Boston: Jones and Bartlett.

Lebowitz, R. J. 1995. *Plant Biotechnology*. Oxford: Brown.

Mauseth, J. D. 1995. *Botany: An Introduction to Plant Biology*. Philadelphia: Saunders.

Electronic resources:

Maas, K. "Plant Hormones and Growth Regulatory Substances." <http://www.plant-hormones.bbsrc.ac.uk/education/Kenhp.htm>

Mather, D., McGill University. "Fertility-Regulating Mechanisms in Plants." *Plant Breeding (Course) Homepage.* <http://gnome.agrenv.mcgill.ca/breeding/fertreg.htm>

NSF Center for Biological Timing. "Biotiming Tutorial: Clocks in Other Organisms. Page 2—Plants." *The Center for Biological Timing Home* <http://www.cbt.virginia.edu/tutorial/OTHERORGCLOCKSP2.html>

Schu, B. and J. F R. Ortigão. "Antisense Oligonucleotide Technology and Chemistry." *Interactiva Virtual Laboratory.* <http://caruso.interactiva.de/oligoman/Antisense-Technology.html>

Wulster, G. J., Rutgers Cooperative Extension. "Fuchsia hybrida Flowering." *Rutgers Floriculture Home Page (Floriculture Publications from RCE).* <http://aesop.rutgers.edu/~floriculture/publications/fuchsia.htm>

Stage 1:
THE MICROBIAL CLEANUP BRIGADE
Deborah E. Allen

••

Daniel DeWitt has just returned to his hometown in Chester County, Pennsylvania, after nearly 20 years' absence. Although he swore he'd never do it, the prospect of assuming a partnership in a local company, BioDecon, Inc., has lured him back. BioDecon designs and implements bioremediation techniques for soils contaminated with industrial by-products, and specializes in biostimulation, or in situ enhancement of the growth of soil microorganisms capable of degrading the toxic wastes. The present owner (Charles MacGregor) wants to use Dan and his reputation to draw much-needed business into the firm.

Several months after Dan's return, Lisa Melrose, a reporter from a local newspaper, receives a call.

"Ya familiar with that place up on Stoltzfus Road—ya know, Fox Hunter's Crossing?" asks the caller (who refuses to identify himself).

"Yes," Lisa replies, wondering how long before she'll be able to politely detach herself from the conversation and head off for lunch.

"Ah, yeah. I'm calling' cause I heard a family trying ta sell just got a negative well report back."

"And why should I be interested in that?" the weary and hungry reporter wonders aloud.

"Ah, yeah. Well, the story is the water's fulla bad stuff—the kinda stuff that mighta leaked out of the below-ground tanks from that plant that's up the road apiece."

The reporter's ears perked up. "And how did you find out about this?" she asks just a little too late. The anonymous source (Deep Throat II?) has just hung up.

Lisa does a little research, and finds out that the story is true. Just as the developer of Fox Hunter's Crossing is beginning to sell homes under Phase II construction, some of the homeowners in Phase I have discovered that their well water contains levels of chlorinated hydrocarbons (trichloroethylene [TCE], chloroform, and methylene chloride) that exceed the maximum contaminant levels established and enforced by the Environmental Protection Agency under the Safe Drinking Water Act. Lisa and her newspaper have a field day (and Lisa gets that long awaited job at a big city newspaper).

And back to Dan....Several months after the news hits the headlines, BioDecon receives a call from an engineering firm. The firm has verified that there is local groundwater contamination (that has seeped into an aquifer tapped by some of the wells; this area does not have a public water and sewer system). It has also found several fairly well delineated areas of soil contaminated with a mix of chlorinated hydrocarbons (primarily TCE) and various aromatic compounds (including tolulene and benzene) near the site where an old refrigerator manufacturing plant had located its underground storage tanks—just a few hundred feet up the road from some of the Fox Hunter's Crossing homesites. The firm also reports soil characteristics that range from silty to clayey, depending on the site.

The engineering firm would like to subcontract with BioDecon to conduct feasibility studies to determine whether in situ biostimulation techniques could be the way to go with the cleanup.

"If you're interested, we need you to commit to the project by the end of the week, and send someone over to the site Monday to collect some soil samples from the contaminated sites," a representative from the firm tells Dan.

Dan and his partner Chuck decide to accept the terms of the contract, so Dan sits down to draft a protocol for what to do with the soil samples when one of the firm's research associates brings them back to BioDecon's laboratory facility on Monday.

Questions to ponder:

- Could leakage from the underground fuel storage tanks account for the presence of the particular contaminate compounds that the engineering firm has identified?
- What evidence does Dan need to collect from these soil samples to make a case that these sites could best be cleaned up by biostimulation? (Note: Information in the next stages will provide a partial answer to this question, so don't read ahead if you'd like to try to figure it out for yourself.)

Stage 2:
THE MICROBIAL CLEANUP BRIGADE

. . . .

Today, Dan is analyzing the initial results of laboratory studies that his research associates have performed on soil samples collected from each contamination site. He is reviewing a brief summary plus a table (Table 1 in Part 1 of the report), then some graphs (Figure 1 in Part 2 of the report) showing the results of community respiration studies of soil samples taken from two of the contamination sites. He still hasn't received the results of chemical analyses of soil samples from each site (prepared as reported for the community respiration studies in Part 2, then spiked with test hydrocarbons and incubated for 12 days). These last studies were conducted in an attempt to confirm the results provided in the report—should he take a chance and make a positive recommendation without the reports from the chemical analysis?

Dan iss feeling the pressure. All parties are more than anxious to pro-ceed, including the developer, a local citizens group (The White Clay Creek Conservancy), Vandelay Industries (the Delaware holding company for the refrigerator manufacturing firm, which has since relocated to another state), and especially the homeowners. And, to top it off, his partner Chuck is breathing down his neck to get a decision by the end of the day. The engineering firm is also anxious to hear whether BioDecon is ready to move on to the next phase of the feasibility study.

"I'm not sure why everyone is in such a rush," Dan muses. "Even if cleanup of the sites doesn't get started for another few months, it's bound to be finished long before the lawyers have resolved who has to pay for it. Oh, well, I had better review the report again. Knowing Chuck, he'll be sticking his head in the door before the hour is up, and I wouldn't want him to catch me daydreaming."

Table 1. Utilization of Hydrocarbons by the Five Selected Isolates

Strain #	TCE	DCE	Toluene	BTX
XF 2A	+	+	+	+
XF 2B	–	–	+	+
XF 8	–	+	+	+
XF 10	–	–	+	+
XF 32	+	+	–	–

Part 1: Summary of Feasibility Study

Soil samples were pulverized, passed through a 1 mm sieve, then 5 grams of each sample were added to separate flasks containing 25 ml of nutrient broth. After 3 to 5 days at room temperature, the samples showing evidence of bacterial growth were centrifuged. The supernatant was added to minimal media containing 100 to 400 mg/L of one of the test hydrocarbons (TCE, dichloroethylene [DCE], toluene, or a mix of benzene, toluene, and xylene [BTX]). Bacterial growth was determined turbidometrically. After repeated streaking of selected colonies onto nutrient agar plates, 45 strains of bacteria were identified and isolated. Five of the isolates proved suitable for follow-up studies that tested their ability to degrade hydrocarbons. Data from these 5 strains are presented in Table 1.

Strains were grown on hydrocarbon substrates for 5 days; the cultures were checked for reduction in the amount of hydrocarbon present, with a "+" indicating reduction, and a "–" indicating no reduction.

Part 2 - Graphs of Community Respiration for Two Sites

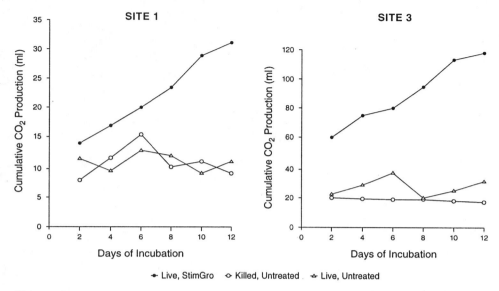

Figure 1. Community respiration of soil samples (prepared as described in Part 1) was determined by measuring CO_2 evolution in biometer flasks. Samples were spiked with several test compounds typical of those found in the contamination sites. For each site, respiration was measured in samples treated with sodium azide to kill resident microorganisms, in untreated samples, and in samples treated with BioDecon's patented additive (©StimGro). Data from only two of the four contamination sites are shown.

Questions to ponder:

- What types of microorganisms found in soil might be suitable candidates for bioremediation at a site contaminated by a mix of organic compounds such as these?
- Why did Dan ask that community respiration be measured?
- What's in StimGro®? What are the possibilities?
- Is in situ biostimulation still in the running as a good way to clean up both sites? If Dan decides that this is so, how will the engineering firm deliver BioDecon's "treatment" to the actual sites of contamination?

Stage 3:

THE MICROBIAL CLEANUP BRIGADE

••

Dan and Chuck agree to give their firm's research associates the green light to spend the next several weeks refining the optimum chemical modifications for enhancement of microbial growth and activity at the various sites. While they are doing this work, additional studies by the engineering firm suggest that at some of the sites, physical alterations of the clayey soil environment will be necessary for success of any bioremediation procedure. In addition, the report for the chemical analysis of compound loss shows that levels of the test hydrocarbons in the soil samples did in fact decrease over the 12 days of incubation.

Dan therefore feels confident reporting a positive outcome to the feasibility studies at a subsequent meeting with the management teams from BioDecon, the engineering firm, and the president of the homeowner's association (which has decided, in a controversial meeting, to assess each Phase I and II homeowner a fee to foot the bill for the feasibility study until liability can be resolved by the courts).

Dan and his partner leave the meeting thinking that they have provided solid evidence for the potential effectiveness of a biostimulation approach to cleanup of the plant. Therefore, they are disappointed to hear several days later that the engineering firm has recommended that a combination of excavation, land farming, and extraction wells will be the most effective cleanup for these relatively shallow contamination sites.

Dan is not the type to let it go at that. He begins to plan another proposal to make to the engineering firm that may allow his firm's services to still be in the running for this potentially lucrative contract.

"We could still be an important part of the monitoring of contamination levels throughout the cleanup process," he says to Chuck. "It might be a way for us to salvage our presence on this project, and, buddy ole pal, we need this contract badly."

Questions to ponder:

- Why are the studies performed by the chemical laboratory considered to be confirmatory?
- If you were the homeowners or the developers, would you be satisfied with the approach recommended by the engineering firm?
- Who should pay for the cost of the cleanup?
- BioDecon uses biostimulation, but other firms use a technique called bioaugmentation. This involves addition of selected nonresident microbes, often genetically engineered, to the cleanup site. What are the advantages and limitations of this alternate approach?

Topics Introduced by the Problem:
THE MICROBIAL CLEANUP BRIGADE

....

Basic Background Topics

(Your textbook can be consulted for information on these.)
- The diversity of organisms found in soil microbial communities
- The different modes of nutrition used by soil microorganisms
- Techniques commonly used to isolate bacterial strains, and to assess their growth and activity
- A review of central metabolic pathways
- Some of the basic features of good experimental design

More Advanced Topics

(You may need to consult additional resources for information on these.)
- An introduction to a technology used to clean up toxic waste sites, including Superfund sites
- How microorganisms metabolize less common organic substrates
- Environmental conditions that optimize the growth of soil microorganisms

Additional Resources for Researching Problem Content:
THE MICROBIAL CLEANUP BRIGADE

Books and articles:

Alexander, M. 1994. *Biodegradation and Bioremediation.* San Diego: Academic Press.

Atlas, R. M., and C. E. Cerniglia. 1995. "Bioremediation of Petroleum Pollutants." *Bioscience* 45:332–339.

Baker, K. S., and D. S. Herson, editors. 1994. *Bioremediation.* New York: McGraw-Hill.

Crawford, R. L., and D. L. Crawford, editors. 1996. *Bioremediation: Principles and Applications.* New York: Cambridge University Press.

Fry, J. C., editor. 1992. *Microbial Control of Pollution.* New York: Cambridge University Press.

King, R. B., G. M. Long, and J. K. Sheldon. 1998. *Practical Environmental Bioremediation: The Field Guide.* Boca Raton, FL: Lewis Publishers.

Schroeder, E. D. 1997. *Bioremediation Principles.* New York: McGraw-Hill.

Snyder, J. D. 1993. "Off-the-Shelf Bugs Hungrily Gobble Our Nastiest Pollutants." *Smithsonian,* 24:66–73.

Stover, D., and R. A. Marini. 1992. "Toxic Avengers." *Popular Science* 241:70–75.

Young, L. Y., and C. E. Cerniglia, editors. 1995. *Microbial Transformation and Degradation of Toxic Organic Chemicals.* New York: Wiley-Liss.

Additional Resources for Researching Problem Content:
THE MICROBIAL CLEANUP BRIGADE

....

Electronic resources:

Bioremediation Resources.
<http://www.nalusda.gov/bic/Biorem/biorem.htm>

Environmental Protection Agency. *Envirofacts Warehouse.*
<http://www.epa.gov/enviro/index_java.html>

Errampalli, D., University of Guelph, and A. R. Piggott, National Water Research Institute. *Bioremediation Resources on the Internet.*
<http://gw2.cciw.ca/internet/bioremediation/>

Ferreira, E. C., Departamento de Engenharia Biológica. Universidade do Minho. EnviroInfo: *Environmental Information Sources.*
<http://www.deb.uminho.pt/fontes/enviroinfo/enviroinfo.htm>

Indiana Institute for Molecular and Cellular Biology, Indiana University. *Biotech Science.* <http://biotech.chem.indiana.edu/lib/orgstrain.html>

Pallen, M. *The Microbial Underground's Guide to Microbiology on the Net.*
<http://www.lsumc.edu/campus/micr/mirror/public_html/microbio.html>

U.S. Environmental Protection Agency. Technology Innovation Office. *Clu-In: Hazardous Waste Clean-Up Information.* <http://clu-in.com/>

U.S. Geological Survey. *Bioremediation: Nature's Way to a Cleaner Environment*
<http://h2o.usgs.gov/public/wid/html/bioremed.html>

U.S. Geological Survey. *Toxic Substances Hydrology Program.*
<http://www.vares.er.usgs.gov/toxics/>

Stage 1:

WHERE HAVE ALL THE FROGGIES GONE?

Deborah E. Allen

Anecdotes about missing frogs ("They were all over the place when I hiked up here five years ago. Where did they go?") started being told in the 1970s, and were relatively easy to dismiss as coincidence. By 1990 the dwindling number of amphibians, particularly in higher altitude populations, began to be acknowledged as being a global phenomenon. Is it possible that amphibians, who have survived hundreds of million of years worth of environmental catastrophes (both large and small), might disappear from the planet?

In late spring of 1998, the situation was serious enough to prompt the National Science Foundation (NSF) to hold a two-day "Workshop on Amphibian Population Decline." An advisory posted on the NSF Web site (<http://www.nsf.gov/pubs/1998/pa9812/pa9812.txt>) alerted the media to the upcoming workshop:

"Increasing Threat of Extinction for Amphibians? Scientists to Seek Answers at NSF Workshop

"Where have all the frogs, toads and salamanders gone? The world's leading researchers on amphibian declines will debate that question, and seek explanations for continuing downward trends of some amphibian populations, at a workshop sponsored by the National Science Foundation (NSF).

"Loss of wetland habitat has reduced populations of frogs and toads, and endangered several species of amphibians with restricted ranges, scientists say. Alarming new events have added to this long-term trend, these researchers believe. Frog and toad populations have declined dramatically in the past several years, many in high-altitude locations in the western United States, and in Puerto Rico, Costa Rica, Panama, Colombia and Australia. Studies suggest that the declines may be caused by infections, perhaps promoted by environmental stressors such as synthetic organic compounds like pesticides, metallic contaminants, acid precipitation, UV-B radiation and increased temperatures.

"These issues and some of the latest research on amphibians will be the subject of the NSF workshop. Experts from leading institutions in this field of biology will attempt to answer the question of whether there's any hope of rescuing the frogs, toads and salamanders of the world before it's too late.

| What: | *Workshop on Amphibian Population Declines* |
| Where: | *National Science Foundation (Exhibit Center, First Floor) 4201 Wilson Blvd., Arlington, Virginia (Ballston Metro Stop)"* |

Kate S. Bianna, a faculty member in the school of public health policy at a large western university, has just read this media alert, sent to her over the Internet from the NSF Custom News Service to which she subscribes. "I wonder why they sent me this," she thinks as she scans it.

As fate would have it, a minute later her computer beeps, prompting her to check her e-mail. It's a message from a government agency representative seeking to set up a task force to make formal recommendations to the Department of the Interior on the scope of the amphibian problem and on possible courses of action. Will Kate serve as a member? (An official letter will follow via snailmail.) Kate quickly taps out a reply, requesting a day or two to think it over.

"I don't know much about frogs, toads and salamanders," she muses while logging out. "But I do know one thing—they come in all shapes and sizes and live just about any place where it's the slightest bit wet. How do they expect a task force to

sort out all of the information on why they seem to be on the decline? About the only thing I can think of right now is that they must be the victims of habitat destruction or disease, just like most other threatened species. Isn't it just that simple? And, do I really care about frogs enough to spend so much time away from my work plotting about their fate? What could a public health expert like me contribute to a task force like that anyway?"

Questions to ponder:

- Should a public health expert like Kate care enough to devote her time to the task force's activities? Why might her insights be needed?
- Could Kate be right? Can the decline in amphibian numbers be attributed mostly to man's destruction of their habitat and to disease? Or, are amphibians somehow the mine canaries of an overall unhealthy state of the planet?
- What is it about amphibians that might make them particularly susceptible to adverse environmental conditions?
- After reviewing the evidence on why amphibian populations are on the decline worldwide, what recommendations do you think the task force should make?

Topics Introduced by the Problem:
WHERE HAVE ALL THE FROGGIES GONE?

Basic Background Topics in Biology
(Your textbook can be consulted for information on these.)
* An introduction to amphibian classification and biology
* Ozone depletion and increased UV-B exposure

More Advanced Topics
(You may want to consult additional resources for information on these.)
* Additional information on amphibian habitats, behavior, diseases, and physiology
* Possible causes of decline in amphibian numbers
* Why species become extinct
* Environmental policy making

Additional Resources for Researching Problem Content:
WHERE HAVE ALL THE FROGGIES GONE?

....

Articles and books:

Blaustein, A. R., and D. B. Wake. 1995. "The Puzzle of Declining Amphibian Populations." *Scientific American* 272:52–57.

Duellman, W. E., and L. Trueb. 1994. *Biology of Amphibians*. Baltimore: Johns Hopkins University Press.

Feder, M. E., and W. W. Burggren, editors. 1992. *Environmental Physiology of the Amphibians*. Chicago: University of Chicago Press.

Gannon, R. 1997. "Frogs in Peril." *Popular Science* 251:84–88.

Heyer, W. R., M. A. Donnelly, R. W. McDiarmid, L.-A. C. Hayek, and M. S. Foster, editors. 1993. *Measuring and Monitoring Biological Diversity—Standard Methods for Amphibians*. Washington, DC: Smithsonian Institution Press.

Luoma, J. R. 1997. "Vanishing Frogs." *Audubon* 99:60–69.

Phillips, K. 1995. "Sun Blasted Frogs." *Discover* 16:76.

Stebbins, R. C., and N. W. Cohen. 1995. *A Natural History of Amphibians*. Princeton, NJ: Princeton University Press.

Electronic resources:

Davidson, C., C. Gregory, L. Hansen, A. Lind, M. Pitkin, and B. Shafer, University of California at Davis. *Westward Frog.*
<http://ice.ucdavis.edu/Toads/herp.html>

Loosemore, S. "The Scientific Amphibian." *The Froggy Page.*
<http://www.frog.simplenet.com/froggy>

Monds, S., Environment Canada. *The Herptox Page.*
<http://www.cciw.ca/greenlane/herptox/>

U. S. Geological Survey. *North American Amphibian Monitoring Program.*
<http://www.im.nbs.gov/amphibs.html>

Stage 1: The Gypsy Caterpillar Invasion

CAMPING WITH CATERPILLARS

Harold B. White

The faint patter of fecal pellets hitting the dry leaves sounded like a steady light rain despite the bright sunshine that filtered through what remained of the shredded oak leaves. There were gypsy moth caterpillars everywhere—thousands, even millions, of them. During the day they crowded into the dark corners and cracks of the cabin eaves and tree trunks. What a contrast to the lush greenery that had greeted Bob June a year ago when he had arrived for his first summer as a counselor at Camp Calosoma.

As he surveyed the scene, Bob noticed that the devastation had a fairly definite boundary and occupied a roughly circular area a few hundred yards across. Parts of Camp Calosoma still had dense shade. Interestingly, within the heavily infested area there were some plants such as tulip poplar and laurel, that seemed to flourish.

Mr. Kalmia, the camp director, met with the counselors that evening to answer questions and discuss plans for the summer. "No, we are not going to spray insecticides." "Yes, some campers might get a rash from handling the caterpillars." "No, they are not poisonous." "Yes, the infesta-

tion could very well become the theme of this year's nature program." He went on to philosophize about how things that seem to be disasters can be opportunities and that the "Invasion of the Gypsy Moths" was an enormous opportunity. "How we deal in life with situations such as this reflects our character. Whether it be gypsy moths, zebra mussels, dandelions, kudzu, the chestnut blight, English sparrows, killer bees, or smallpox, humans have intentionally or accidentally introduced exotic species into their environment and occasionally lived to regret it. There are important lessons to be learned. Please think about this and come up with ideas for a creative summer program."

Bob suppressed his first reaction to find another summer job and began brainstorming with the other counselors.

Questions to ponder:

- What do you know about the life cycle of moths and butterflies?
- How is it possible for one species to have a sudden population explosion?
- How can plants defend themselves against caterpillars?

Stage 2:   *The Lights Go on for Bob*
CAMPING WITH CATERPILLARS

It was well after dark when the counselors' meeting ended. As Bob headed back to his bunk for the night, the insects around a light distracted him momentarily. In particular, he noticed several large iridescent green beetles commonly known as caterpillar hunters. It suddenly dawned on him,."These beetles must be feasting on gypsy moth caterpillars. And the cuckoos I heard this afternoon—they have an appetite for caterpillars as well. We humans must be oblivious to all sorts of things happening around here due to the gypsy moth infestation. Maybe exploring all the links between gypsy moths and other organisms could be the kind of opportunistic program theme for the campers that Mr. Kalmia would like."

Bob's clock read 2 A.M. He couldn't get to sleep. His mind still was playing. "Though harmless to us, how different the green beetles would look through caterpillar eyes. The campers could identify with that by imagining themselves encountering a tiger. Maybe a survival game, "Checks and Balances," involving various trees, birds, insects, and humans, could be created. What about having the campers see if the insects attracted to light were different in the infested and noninfected parts of the camp?" The possibilities seemed to multiply.

Questions to ponder:

- Imagine you are a gypsy moth caterpillar. What factors will affect your survival? Which of these are beyond your control? Which might you influence in your favor? Most of the 9- and 10-year-old

Stage 3: A Mystery Killer
CAMPING WITH CATERPILLARS

boys assigned to Bob's cabin got interested in the caterpillars as a part of the "Camping with Caterpillars" theme. His group had decided to see if starving caterpillars would eat sassafras, laurel, or tulip poplar leaves. Another group wanted to see how many caterpillars the caterpillar hunters could eat. Yet another group tried different ways to protect trees at the edge of the gypsy moth infestation. A few days into the second week of camp, after a few damp days, Joey, a camper who got up early for a walk every day, woke everybody shouting, "They're all dead! They're all dead!"

Sure enough, dead caterpillars by the thousands hung limply on the tree trunks. Only a few survived. Interestingly, all of those kept in jars by the campers survived. Other insects in the area seemed unaffected.

Questions to ponder:

- What might have caused the caterpillars to die? Are any possibilities ruled out by the information provided?
- If you were Bob, how could you integrate this event into the summer program?

Basic Background Topics

Topics Introduced by the Problem:
CAMPING WITH CATERPILLARS

(Your textbook can be consulted for information on these.)
- Factors that influence population size
- R- versus K-selection
- Characteristics of a temperate deciduous biome
- Trophic levels
- General stages in insect life cycles
- General aspects of insect classification

More Advanced Topics
(You may want to consult additional resources for information on these.)
- Defenses of tree against insect predators
- Allelochemicals
- The oak-hardwood forest ecosystem
- The life cycle and life history of the gypsy moth

Additional Resources for Researching Problem Content:
CAMPING WITH CATERPILLARS

• •

Articles and books:

Elkinton, J. S., and A. M. Liebhold. 1990. "Population Dynamics of Gypsy Moth in North America." *Annual Review of Entomology* 35: 571–596.

Herrick, O. W., and D. A. Gansner. 1987. "Mortality Risks for Forest Trees Threatened with Gypsy Moth Infestation." *USDA Forest Service Reseach Note* NE-338.

Montgomery, M. E., and W. E. Wallner. 1988. "The Gypsy Moth: A Westward Migrant." In *Dynamics of Forest Insect Populations*, A.A. Berryman, ed. New York: Plenum.

Romoser, W. S., and J. G. Stoffolano Jr. 1998. *The Science of Entomology*. Boston: WCB/McGraw Hill.

Weseloh, R. M. 1985. Predation by *Calosoma sycophanta* (Coleoptera: Carabidae): Evidence for a Large Impact on Gypsy Moth, *Lymantria dispar* (Lepidoptera: Lymantriidae) Pupae." *Canadian Entomology*117:1117–1126.

Electronic resources:

National Park Service. Western Regional Server of National Integrated Pest Management (IPM) Network. "Gypsy Moth." *Integrated Pest Management Manual.* <http://www.colostate.edu/Depts/IPM/natparks/gypsymth.html>

Stage 1: *The Beleaguered Scientific Expert*

ON THE ROAD TO EXTINCTION?

Richard S. Donham and Deborah E. Allen

Dr. Clarence Leopold is chief of the Endangered Species Section for Fisheries and Wildlife and has been occasionally asked by the Secretary of the Interior to provide scientific input to an Executive Branch committee that considers proposed domestic and international economic aid. Today, the committee is meeting to consider a proposal for agricultural aid to Sub-Saharan Africa that may impact on the wildlife of the area. As he awaits his turn for presentation, Clarence thinks: "This is not what I had in mind when I went to veterinary school." He shifts his gaze around the table, and hears a committee member from the State Department conclude: "This will prevent widespread famine and promote regional stability."

Clarence then presents his carefully reasoned presentation of the danger of agricultural development to the wildlife, emphasizing in particular the devastating potential on certain "charismatic" endangered species.

During the open discussion period following, Claudia Ostrinia, from Agriculture, points out: "Dr. Leopold, I remember several years ago your telling this committee that the cheetah was in trouble because of 'population bottlenecks,' and something about genetic heterozygosity. My question for you is: if the cheetah's problems are not ecological, and the animal is doomed anyway, why not spend money on a project that may help humans?"

Hobart Strongylus from Interior chimes in: "Supposing, for the moment, that the cheetah is not in genetic trouble but, because the population is decreasing, maybe we don't know the reason for its impending demise. If this is so, reasonable agricultural development may not affect them at all." ("Incompetent traitor," thinks Clarence.)

There is further discussion, mostly centered on the meaning of genetics to the survival of the cheetah and other endangered species. Finally, Raymond Carson from Education inserts, "I move that, in the interest of pushing this discussion toward a conclusion, Dr. Leopold prepare for consideration at the next meeting a clear and in-depth explanation of the significance of these genetic arguments to the survival of the species." After discussion, the motion is carried.

Sitting later at his desk, Clarence gazes at his notes of the meeting, including some terminology that he used to further his position, which, inexplicably, seemed to make the opposition even more entrenched: genomic variation, mitochondrial DNA, and immunological incompetance. Plain enough terms he thought, and he couldn't quite figure out why the committee members should be so hostile.

Questions to ponder:

- What is a population bottleneck and what may be the result on the genetics of a species? Is there evidence that the cheetah has experienced a population bottleneck? Are there other examples, and what has happened to these organisms?
- How much variation is there in a species of animals, that is, how is it measured, and what is the significance?

Stage 2: . *Another not-so-satisfying meeting*
ON THE ROAD TO EXTINCTION?

At the next meeting of the committee, the discussion continues, and it is clear that not all of the members have found Clarence's report persuasive. In addition, some have been doing independent investigations.

"Clarence" (this is Lauren Seals, the Committee Chair, and Clarence is not sure he likes this progression toward familiarity), "in your report you say that the cheetah has a lot of homozygosity in the genome. I believe you say that they are all about as closely related as inbred strains of mice. Then you argue that, as this example illustrates, being inbred doesn't necessarily prevent an organism from being successful. However, I have also read that sperm collected from cheetahs are deficient, that fertility is low, and that they have suffered mortality from a disease that other big cats tolerate quite well. Which is the case: is homozygosity not critical, as you argue, or are these creatures, beautiful they may be, on a one-way road to extinction? If the latter is true, I don't see why we should worry too much about agricultural development perhaps resulting in their demise a few years earlier than if we try to save them!"

"I tend to agree," chimes in Laurel Zia from Agriculture. "After all, extinction is a natural process, we all know what happened to dinosaurs and saber-tooth tigers."

"The chair is correct about captive cheetahs suffering from reproductive deficits, but the situation in Africa is not so dismal. Those deaths you referred to were in a captive group in Oregon, and that may have been a husbandry problem. As to extinction being a natural process, I agree with my colleague from Agriculture, but the rate of extinction occurring now may be unprecedented, and the increases in extinction rates since 1800 are thought to almost exactly parallel the exponential increase in human population." Clarence sits back in his chair, pleased with his rejoinder.

"Dr. Leopold," rumbles Oliver Murie from Interior, "I seem to recall from my college biology that during sperm production variation was introduced into our genetic makeup, and so I am not simply a mix of my father's and mother's traits, but am unique, as we all are. If this is so, perhaps the fact that cheetahs have few and misshapen sperm means that the whole process of sperm production is messed up in these cats and that the lack of genetic variation is a reflection of this, not the other way around."

Oliver continues: "Since we may never resolve some of these issues, I suggest a simple experiment is needed—how about moving a group of cheetahs to a new area and seeing how they do?"

Questions to ponder:

- What is the process Oliver is referring to, and is he right in suggesting that genetic variation is introduced during this process?
- The Committee doesn't seem to be convinced of the power of Clarence's arguments or his report. What should he do to change the doubters' opinions?
- What hypotheses would be tested by moving cheetahs to a new location? Where would you move them?

Topics Introduced by the Problem:
ON THE ROAD TO EXTINCTION?

· ·

Basic Background Topics in Biology
(Your textbook can be consulted for information on these.)
• Genomic variation (polymorphisms)
• Nuclear and mitochondrial DNA
• Meiosis and gamete formation
• Mechanisms for introducing genetic variability
• Basic mechanisms of evolution

More Advanced Topics
(You may want to consult additional resources for information on these.)
• Rates of evolution
• Major histocompatibility complex (MHC)
• Immunological competence
• Polymorphisms in length of restriction enzyme fragments
• Convergence between molecular genetics and conservation biology

Additional Resources for Researching Problem Content:
ON THE ROAD TO EXTINCTION?

. . . .

Articles and books

Caro, T. M., and M. Karen Laurenson. 1994. "Ecological and Genetic Factors in Conservation: A Cautionary Tale." *Science* 263:485–486.

Darnell, J. E., and H. Lodish. 1995. *Molecular Cell Biology.* New York: W.H. Freeman & Company.

O'Brien, S. J., D. E. Wildt, and M. Bush. 1986. "The Cheetah in Genetic Peril." *Scientific American* 254:84–92.

O'Brien, S. J. 1994. "A Role for Molecular Genetics in Biological Conservation." *Proc. Natl. Acad. Sci. USA* 91:5748–5755.

Pennisi, E. 1993. "Cheetah Countdown." *Science News* 144:200–201.

Electronic resources:

American Zoo and Aquarium Association. *Cheetah Fact Sheet.*
<http://www.aza.org/aza/ssp/cheetah.html>

Cheetah Conservation Fund. *Introduction to the Cheetah.*
<http://www.cheetah.org/>

The Cheetah Spot. *Facts, Pictures, Sounds and Links.*
<http://www.neocomm.net/~eadams/cheetah.html>

A BIRDSONG TRILOGY

Deborah E. Allen

....

Where do biologists get the hypotheses (or possible answers to questions) they design their experiments to test? Many of the most intriguing hypotheses stem from curiosity about events or living things that biologists observe in the natural world (or in the simulated world within the laboratory), or are generated by the rarer and lucky biologists who can see the familiar world with a new eye. They are possible and plausible explanations for why these observed phenomena exist, or for how they might work. Ideally experiments are designed to see if there is clear evidence for supporting or rejecting a hypothesis (or for a "yes" or "no" answer to a question), but often experiments become a source of new hypotheses because they inevitably open up new windows on the living world.

To be useful to a biologist, regardless of the source, a hypothesis needs to be testable—that is, there has to be a reliable, repeatable way of finding the clear-cut answer that the biologist seeks. A good way to find out if one is testable is to try to make a prediction from it. A person trying to make a prediction is thinking: "If my hypothesis is true, what does that imply about the various situations in which I think this hypothesis can be implicated? What other things will I

observe as a consequence, if this is true?" If it is tough to come up even one prediction, chances are that the hypothesis is either poorly stated or difficult to subject to a test, and therefore not yet of much practical use.

The scenario below (*Why Do Birds Sing?*) and the questions and statements that follow are designed to give you a better idea of the link between a hypothesis and a prediction, and—to take it one step further—to see how a prediction can give a biologist a good place to start in designing the experiment best suited to testing the hypothesis. The scenario begins with the song of a Northern cardinal, a bird that is easy to spot at most times of the year, and in many habitats (including suburban yards) east of the Rockies. If you are reading this in a classroom in another part of the country, feel free to write a local bird into the scenario. Many birds sing, and those that do tend to do more singing at certain times of the year.

Why Do Birds Sing?

When you come to class some morning this spring, you may hear male cardinals (those bright red birds) singing, "Cheer! Cheer! Cheer!"; house finches warbling from somewhere in the trees overhead; or mock-

ingbirds running through their reper-
toire of mimicked songs (their version
of songs they hear other species sing).
In the woods surrounding a nearby
creek, you might hear, if you took a
morning walk, 20 or 30 distinctly different songs, although you might see
only a few of the birds singing them.
Why do birds sing? Biologists have
observed that most singing occurs in
the spring, although there are some
birds that sing at other times of the
year. Often (although not always) the
male does the singing.

Some biologists have hypothe-
sized that singing is a way that male
birds notify other male birds to keep
out of their territory. If this is true,
what aspects of bird behavior would
you predict to occur as a consequence? One prediction that comes to
mind is that if this is so, singing
should occur only when a male bird is
on his territory. A second prediction
is that singing should then occur
whether or not a female bird is
around because the hypothesis states
that it is a response to a competitor
for the territory, not to a welcome
addition.

Another possible explanation for
singing is that it is what male birds do
to attract the attention of females. A
prediction that could be made from
this hypothesis is that if this is true,
then the presence of a female should
stimulate a male to sing.

- Are there any other plausible
  hypotheses (educated guesses) that
  you can make about why birds sing
  from the information given in the
  scenario?

This first exercise might have
given you some ideas about why predictions are so useful in designing
experiments. Armed with these predictions, the interested ornithologist
has a basic premise around which an
experiment can be designed. He or
she would then need to make some
clever decisions about how to map a
bird's territory, how to tell a song
from other sounds that a bird makes,
when to make the observations, how
many birds to observe, how to tell if a
bird is male or female, and how to
keep track of individual birds. The
most difficult part, however, would
be designing an experiment that eliminated all of the other reasonable
alternative hypotheses from consideration, providing clear-cut evidence
that could lead to an acceptance or
rejection of the hypothesis being tested.

The two scenarios on the next
pages (*What's the Attraction Here?*
and *How Does a Turkey Vulture Find
a Road Kill?*) will give you some
additional opportunities to practice
making hypotheses and predictions,
and to design experiments to test
your hypotheses.

What's the Attraction Here?
A BIRDSONG TRILOGY

The history of cattle egrets in this country is a true American success story. Before the 1880s, an American would have to travel to Africa or southern Eurasia to see a cattle egret. Before 1942, a North American would have to hop a plane to Rio, or to some other location in South or Central America to get a glimpse of one (sometime in the late 1800s, egrets crossed the Atlantic to take up residence on these new continents). In 1952, the first "blessed event" (purely from the egret perspective) was recorded in North America—nests full of egret hatchlings were found in Florida. Within 10 years of this first recorded breeding, a small group of about seven pairs of Florida pioneers had become a colony of 4000 egrets. By 1972, cattle egrets had become the most plentiful egret in North America, found nesting in all but a dozen or so states. Now, if you took a summer walk in the countryside in any state, you would probably see at least a few of these medium-sized white birds strolling around in the pastures, often around the feet of cattle.

Although their white color is a standout, another distinctive aspect of these birds is the way they poke around in the vegetation, then abruptly bring their bills up and jerk their heads back as they swallow the grasshoppers and other field insects that are the main course on the cattle egret menu.

African cattle egrets associate with buffalo, elephants, and other large grazing animals. In North America, they like to hang out with domesticated cows. Usually, if egrets are present in a field, some egrets will be found near the cattle, while others can be observed moving around at some distance away from the cows.

- Why are cattle egrets found around cattle? Give at least two plausible explanations (hypotheses). (*Note: The next page provides an answer to this question, so don't look ahead if you would like to figure it out for yourself.*)

One hypothesis that might explain the egrets' behavior is that associating with cows is a way to get the cows to do some of the work of finding food. The cows may act like "beaters," stirring up insects during their grazing, and this may make the insects (who often are camouflaged to match the color of the surrounding vegetation) easier targets for the egrets.

- What predictions about cattle egret behavior can you make from this hypothesis? Or, pick a hypothesis that you came up with on the previous page, and decide what predictions you can make from it. (*Note: The next page provides the answer to this question, so don't look ahead if you'd like to figure it out for yourself.*)

One prediction that can be made from the food availability hypothesis is that egrets near the cows should catch more insects than those farther away because it would be easier for the egrets at the feet of cows to spot food.

- If this hypothesis is a testable one, you should be able to use a prediction you made from it as a starting point for designing a experiment. Try this out for the prediction made for the food availability hypothesis. Or, if you prefer, design an experiment to test of one of the alternate hypotheses you came up with in the first stage of this exercise (but don't forget that making a prediction first will help).

Had enough of birds and bird-brained experiments? If not, try your hand at designing an experiment on turkey vultures from the scenario presented on the next page.

Turkey vultures are large dark birds with featherless (bald) red heads (not even the chicks are all that

How Does a Turkey Vulture Find a Road-Kill?
A BIRDSONG TRILOGY

lovable looking), and are common in most of North America. They eat carrion that they find by soaring on updrafts and then slowly dropping down (you've probably seen this scenario in more than one B-grade Western movie). How do they find these delectable morsels? Perhaps they follow the scent of putrefying bodies—a distinctive odor from any perspective. But maybe not—most birds have a poor sense of smell. Maybe they use their vision—the eyesight of most birds is good. The question of how turkey vultures find food should be answerable using the scientific method.

- The discussion above provides two alternate hypotheses for how turkey vultures find food. Are there any others that seem reasonable?
- Pick your favorite hypothesis, and make some predictions that would follow if it is true.
- Use one of your predictions to design a test of the hypothesis, including as many specifics as possible about what you would do, what kind of observations you would make, and how you would record your data.

Additional questions to ponder:

- Now that you have designed some experiments on birds, would you be interested in going out and trying to conduct them?
- In your opinion, should biologists be spending their time investigating questions like these?
- Should taxpayer dollars be spent helping biologists to do experiments that satisfy curiosity about the natural world, but have no immediately identifiable health or economic benefit?

Dr. Richard Donham (who never tires of talking about birds) provided valuable insights on birds and bird lore as this problem was developed; his assistance is gratefully acknowledged.

Basic Background Topics

Topics Introduced by the Problem:
A BIRDSONG TRILOGY

(Your textbook can be consulted for information on these.)
- The scientific method—how scientists formulate hypotheses and design experiments to test them

Additional Resources for Researching Problem Content:
A BIRDSONG TRILOGY

Books:

Farrand, J., Jr. 1984. *The Audubon Society Master Guide to Birding.* New York: Alfred A. Knopf.

Grubb, T. C., Jr. 1986. *Beyond Birding: Field Projects for Inquisitive Birders.* Pacific Grove, CA: Boxwood Press. (Chapters 8 and 12 ["Can Turkey Vultures Find Their Way to Food?" and "Why Do Cattle Egrets Associate with Cattle"] provide suggestions on how to test the hypotheses you may have posed. This book was used to develop the two problem scenarios about these birds, and is an excellent source of ideas for field experiments.)

Kaufman, K. 1996. *Lives of North American Birds.* Boston: Houghton Mifflin.

Stokes, D., and L. Stokes. 1996. *Stokes Field Guide to Birds.* New York: Little, Brown and Company.

Welty, J. C. 1975. *The Life of Birds.* Philadelphia: W.B. Saunders.

Electronic resources:

American Birding Association Online. <http://www.americanbirding.org>

Cornell Lab of Ornithology Home Page. <http://birds.cornell.edu>

Siler, J. "Bird Families of the World." *Birding on the Web.*
<http://www-stat.wharton.upenn.edu/~siler/birdframe.html>

Smithsonian Migratory Bird Center Home Page.
<http://www.si.edu/natzoo/zooview/smbc/smbchome.htm>

Swarz, D. *Deb Zone—Home of the Wild Bird Cam.*
<http://host.fptoday.com/debzone>

PART III

FORMS TO HELP YOUR GROUP FUNCTION WELL

Checklist for Problem-Solving

Group Name _____ Date _____

Has today's progress been summarized in a way that everyone understands?

Does everyone understand the importance and relevance of the learning issues discussed or identified today?

Record the learning issues and person(s) assigned to research it.

Learning Issue	Name of Researcher

Has everyone discussed what resources will be used and where to find them?

Checklist for Problem-Solving (page 2)

Group Name _____ Date _____

Have assignments been distributed equally among group members?

Are there areas of misunderstanding that your instructor needs to help clarify?

What aspect did each individual do well today?

What aspects of the group activity need to be improved?

Monitoring Group Function

Group Name _____ Date

Names of Group Members Can Be Written at the Head of the Columns Below

Did He/She:				
Attend Class				
Come Prepared				
Participate in Discussion				
Ask Probing Questions				
Fulfill His/Her Role				
Listen to Others' Views				

Evaluation Form for Individual Effort—Version 1

Your Name _____

Names of Group Members: 1. _____

                                     2. _____

                                     3. _____

                                     4. _____

Please use the following form to assess the contributions of everyone in your group, including yourself. Using the following scale, rate each member of your group (1,2, 3, and 4). Then rate yourself under the column "you" using the following scale: 5 = strongly agree ——> 1 = strongly disagree

	1	2	3	4	you
1. Completed assigned work thoroughly.	___	___	___	___	___
2. Completed assigned work on time.	___	___	___	___	___
3. Contributed relevant information when solving group problem.	___	___	___	___	___
4. Asked questions in a way that promoted clearer understanding of problem.	___	___	___	___	___
5. Answered group members' questions in a way that promoted understanding.	___	___	___	___	___
6. Fulfilled his/her role of responsibility.	___	___	___	___	___
7. Overall, I rate this individual's contribution to our group effort: 5 : excellent ——> 1 : very poor	___	___	___	___	___

Additional comments on individuals:

1.

_____

2.

_____

3.

_____

4.

_____

Evaluation Form for Individual Effort—Version 2

Name of Person
You Are Assessing _____ Your Name _____

Check one box for each assessment item to indicate the extent to which you agree it describes the person you're assessing. Use the space on the next page for specific comments (required if you've checked the strongly disagree, disagree, or strongly agree boxes, with the following exceptions: a strongly agree response for items 1 and 2.)

Group Function	Strongly disagree	Disagree	Weakly disagree	Agree	Strongly agree
1 Does not miss out on group activities by being absent or late	☐ 1	☐ 2	☐ 3	☐ 4	☐ 5
2 Completes all tasks assigned by the group; comes to class prepared	☐ 1	☐ 2	☐ 3	☐ 4	☐ 5
3 Listens to and shows respect for the ideas and opinions of others	☐ 1	☐ 2	☐ 3	☐ 4	☐ 5
4 Does not dominate the discussion	☐ 1	☐ 2	☐ 3	☐ 4	☐ 5
5 Brings new and relevant information to group discussions	☐ 1	☐ 2	☐ 3	☐ 4	☐ 5
6 Reflects and comments on how the contributions of others help with understanding of the problem	☐ 1	☐ 2	☐ 3	☐ 4	☐ 5
7 Asks questions during presentations and discussions that promote clearer and deeper understanding	☐ 1	☐ 2	☐ 3	☐ 4	☐ 5
8 Presents logical arguments and conclusions	☐ 1	☐ 2	☐ 3	☐ 4	☐ 5
9 Communicates ideas and information clearly	☐ 1	☐ 2	☐ 3	☐ 4	☐ 5
10 Helps to identify and implement ways that the group can function better	☐ 1	☐ 2	☐ 3	☐ 4	☐ 5

Evaluation Form for Individual Effort—Version 2 (page 2)

Describe the <u>one</u> aspect of this person's contribution that you think is the most helpful to functioning of your group.

Describe the <u>one</u> way in which you think this person could improve his or her contribution to overall functioning of your group.

Specific comments about responses:

Prompts for Discussion of Group Function

What is the group doing well?

Are all members allowed to initiate and continue discussion?

Are everyone's views listened to respectfully?

Are group assignments being completed to everyone's satisfaction?

Is the group moving at an appropriate pace?

Is the work divided equitably?

Are group members accepting responsibility for learning issues?

What needs to be improved in the group function?

PART IV

SOME HELP WITH SPECIAL ASSIGNMENTS

As you work through some of the problems in this book, your instructor may be interested in using other activities to enhance your understanding of the topics that are being researched or add variety to the groups' discussions. Two of these—jigsaw grouping and concept mapping—are described below.

Jigsaw Grouping

You will be accustomed to discussing learning issues within your group as you work together, doing research, and teaching each other in order to solve a problem. In a jigsaw grouping, you will begin a problem in your group, but will be assigned to form a new group for a limited amount of time so that you can talk with others who are also researching the same learning issues or gathering evidence to support a particular view or opinion. For example, suppose you were beginning to discuss the first stage of the problem, "Human Immunodeficiency Virus and the Health-Care Professional," and your group was listing everything that each student knows about AIDS and HIV, as well as the Kimberley Bergalis case. After identifying the learning issues that arose during your discussion, and attempting to discuss the questions at the end of the first stage, each individual in the group is asked by the instructor to consider the proposition that all health-care professionals undergo mandatory testing for HIV, and if testing positive must reveal the results to all patients. Each individual in the group should choose a point of view for which he or she will become an advocate. For example, some possible options are the following:

- health-care worker in favor of testing and revealing information to patients

- health-care worker opposed to mandatory testing and informing patients

- member of a patient advocate group in favor of testing and informing

- member of a gay activist group opposed to testing and informing

- representative of the insurance industry in favor of the proposition

- member of a citizen's rights group opposed

All students now form new groups that are specific to their point of view; for example, all members of the "citizen's rights group" will meet together. The diagram below shows how this jigsaw group could be formed from the students represented by the number one in each of four different permanent ("home") groups.

Students would then research and discuss a particular point of view, depending on what jigsaw group they have joined. Discussion focuses on issues that affect the point of view. For example, one learning issue for the jigsaw group "health-care workers opposed to mandatory testing and informing" may be, "How many patients have gotten HIV from their health-care worker as opposed to health-care workers who have contracted it from their patients?" Each member of the interest group is assigned learning issues to research, and in the next class session the members share information and construct evidence and argument to support their viewpoint. Then the students return to their permanent ("home") groups to advocate for their positions. The instructor assigns each group to come to consensus on a group position with regard to the proposal,

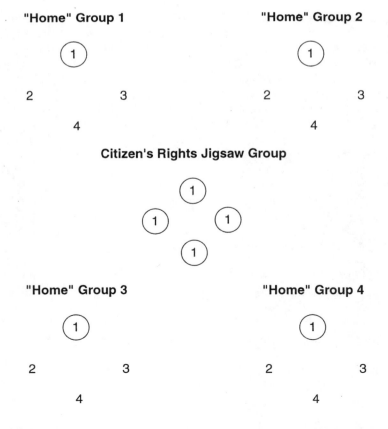

listing the reasons for the group decision. Full class discussion or debate occurs at this point.

A jigsaw grouping can be an effective way to encourage students to fully research one point of view or one learning issue in collaboration with others before sharing that information with their permanent groups.

Concept Mapping

Concept maps are an effective tool for learning, organizing, and retaining concepts. They are also remarkably successful in helping students construct a framework for understanding the connections among ideas, and the hierarchy of operations within that concept. Concept maps are typically constructed in the following manner:

1. Words or a series of words representing key concepts are identified. It is convenient to write these words on "post-it" notes until the map takes its final shape.

2. The words (notes) can be arranged in hierarchies of importance, with more general concepts at the top. The concepts would then get increasingly more specific toward the bottom of the map. Some maps are constructed with the more general (or central) concept in the middle and the more specific ones fanning out from the center.

3. Lines are then drawn between related concepts and a word or phrase that establishes the link between the concepts written above or near each line (these words are referred to as propositional linkages).

4. Crosslinks are then constructed—these are connections between concepts in different areas of the map.

Shown below is an assignment for students as they finish the "Geritol Solution" problem.

Construct a concept map with one of the following titles:
    The light-dependent reactions of photosynthesis
    The light-independent reactions of photosynthesis
    The global carbon cycle
    The flow of energy through the biosphere
The map should contain a portion that clearly indicates how it relates to the "Geritol Solution" problem.

The diagram shown on the next page is a concept map constructed by a typical group of students. Some of the propositional linkages are not complete and some others may not be ones that you would choose to use. As a practice activity, your group may want to construct your own concept map on the flow of energy through the biosphere. How is your map the same or different from the one produced by this group of students?

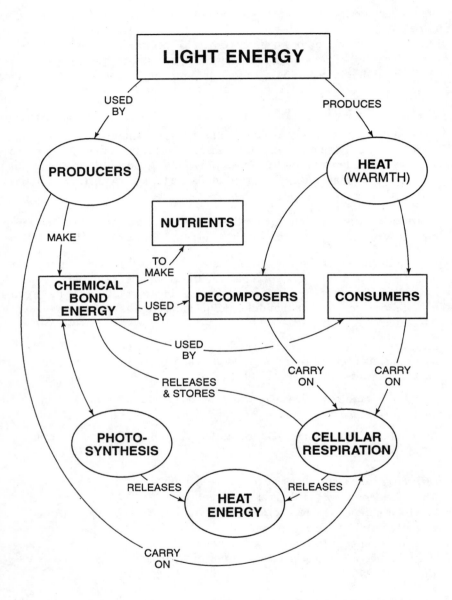

Flow of Energy Through the Biosphere

PART V

BIBLIOGRAPHY

Albanese, M. A., and S. Mitchell. 1993. "Problem-based Learning: A Review of Literature on Its Outcomes and Implementation Issues." *Academic Medicine* 68:52–81.

Bodner, D. 1992. "Why Changing the Curriculum May Not Be Enough." *J. Chem. Ed.* Vol. 69: 186–190.

Boud, D., and G. Feletti. 1996. *The Challenge of Problem-Based Learning*. New York: St. Martin's Press.

Gibbs, G. 1995. *Learning in Teams*. Oxford, England: Oxonian Rewley Press.

Johnson, D. W., R. T. Johnson, and K. A. Smith. 1991. "Cooperative Learning: Increasing College Faculty Instructional Productivity." *ERIC-ASHE Higher Education Report No. 4,* George Washington University.